An Introduction to
Bioenergy

An Introduction to Bioenergy

Nigel G Halford

Rothamsted Research, UK

Imperial College Press

ICP

Published by

Imperial College Press
57 Shelton Street
Covent Garden
London WC2H 9HE

Distributed by

World Scientific Publishing Co. Pte. Ltd.
5 Toh Tuck Link, Singapore 596224
USA office: 27 Warren Street, Suite 401-402, Hackensack, NJ 07601
UK office: 57 Shelton Street, Covent Garden, London WC2H 9HE

Library of Congress Cataloging-in-Publication Data
Halford, Nigel G.
    An introduction to bioenergy / by Nigel G. Halford (Rothamsted Research, UK).
       pages cm
    Includes bibliographical references and index.
    ISBN 978-1-78326-623-4 (hardcover : alk. paper) -- ISBN 978-1-78326-624-1 (pbk. : alk. paper)
    -- ISBN 978-1-78326-625-8 (electronic)
    1. Biomass energy. I. Title.
    TP339.H3477 2014
    333.95'39--dc23
                                2014044354

British Library Cataloguing-in-Publication Data
A catalogue record for this book is available from the British Library.

Typeset by Stallion Press
Email: enquiries@stallionpress.com

CONTENTS

PREFACE

Readers who are old enough to remember the 1970s may well associate that decade with energy crises. Firstly there was the 1973 oil crisis, which started in October of that year when members of the Organization of Arab Petroleum Exporting Countries (OAPEC), which comprised the Arab members of OPEC together with Egypt, Syria and Tunisia, proclaimed an oil embargo. The embargo lasted nearly six months, during which the price of oil quadrupled from US$3 to US$12 per barrel. In the United Kingdom (UK), the effect of the oil crisis was exacerbated by a national miners' strike. It is no exaggeration to say that the combined effect on oil and coal supply crippled the country: from the 1st of January to the 7th of March 1974, much of UK industry was limited to three days of electricity consumption per week, television companies were required to go off the air at 10.30 in the evening and domestic users were subjected to regular power cuts. The crisis brought down the government; indeed, there were two general elections that year.

These events may have occurred four decades ago, but for successive governments since then the 'three-day week', as it was called, has been a reminder of the need to ensure energy security. The energy crises of the 1970s also highlighted the dependence of our way of life on a stable and cheap energy supply; a supply that was and still is heavily dependent on fossil fuels that will eventually run out. Large reserves of coal, oil and gas have been discovered since the 1970s, and the new technique of fracking to extract gas and oil reserves from shale rock has added to those reserves. Nevertheless, the realisation that fossil fuel reserves are finite and alternatives will have to be found has not been forgotten, and 'peak oil', the point at which oil production goes into terminal decline, is now forecast to be reached in the 2020s.

By the turn of this century, a second reason for finding alternatives to fossil fuels had arisen with the realisation that the carbon dioxide released during the burning of fossil fuels is acting as a so-called 'greenhouse gas', preventing thermal energy from escaping the atmosphere and causing global warming and climate change. This coincided with a period when the prices that farmers were receiving for their produce were at historically low levels and many farmers were finding it difficult to make a living. With Brazil having shown, since the 1970s, that it was possible to make ethanol from sugar cane in sufficient quantities to provide a significant proportion of the country's liquid transport fuel, the concept of bioenergy, that is, generating energy from biological sources, including crop products, looked like a win-win option as an alternative to energy from fossil fuels, reducing carbon dioxide emissions, increasing farm prices and diversifying energy generation. This led to political intervention in many countries to promote the development of bioenergy, in the form of liquid transport fuel (biofuel, subdivided into bioethanol and biodiesel); solid fuel, principally for burning to generate electricity (biomass) and renewable natural gas (biogas). As a result, the bioenergy industry has seen huge growth and it now represents a major market not only for established crop products but for a number of plant species not previously considered for cultivation and a variety of waste products.

Perhaps predictably, the notion that bioenergy generation would be a win-win industry has proved to be simplistic and over-optimistic, partly because other events pushed food security to the top of the international agenda. These included severe droughts in Australia and Russia, perhaps a portent of things to come as a result of climate change, and increasing consumption in China,

India and other emerging nations. Using crop products to generate energy, instead of producing food, suddenly became a very controversial issue in a food versus fuel debate. Nevertheless, the industry has continued to develop at a rapid pace, contributing substantially to huge changes in agricultural commodity markets that have occurred over the last decade.

This book is an introduction to the bioenergy industry, describing the feedstocks that are used, including established and potential crops, as well as waste, the production processes, the products, the political interventions to support the industry and the impacts the industry has had on markets. It provides information on how this sector is developing and where it may be headed, and aims to give a balanced view on the arguments for and against the exploitation of different bioenergy sources.

Acknowledgement

The author is supported through the 20:20® Wheat Programme at Rothamsted Research by the Biotechnology and Biological Sciences Research Council of the United Kingdom.

1 INTRODUCTION AND DEFINITIONS

1.1 The Need for Alternative Fuels

Bioenergy can be defined in simple terms as energy generated from materials derived from biological sources. There is nothing new about using biological resources to produce energy: wood has been used to provide energy for warmth and cooking for hundreds of thousands of years, if not longer, and was the most commonly used domestic fuel right up to the 16th century. By historical times, much of the wood was converted to charcoal before use, a process that involves heating under oxygen-limited conditions. Indeed, charcoal production was a huge industry in Europe during the Middle Ages and was responsible for widespread deforestation. Charcoal was displaced by the fossil fuel coal in Europe as the woodlands became depleted and the methods for mining coal from sometimes deep reserves were developed. Nevertheless, the United Nations Food and Agriculture Organisation reported that 1.87 billion cubic metres of roundwood (wood that is trimmed and cut into logs) was used for charcoal production worldwide in 2012. Some of this may have come from sustainable forestry operations, but there is no doubt that the use of wood for charcoal production continues to contribute to deforestation.

Coal and the other fossil fuels, petroleum (oil) and natural gas, have biological origins themselves, of course, being concentrated organic compounds that formed from the remains of plants and animals that lived millions of years ago. However, they are now considered to be mineral rather than biological and therefore do not come within the definition of biofuels. Nevertheless, some commentators prefer to include the term renewable in the definition of bioenergy in order to exclude energy derived from fossil fuels, or to define bioenergy as energy derived from material of recent biological origin.

The modern world is heavily dependent on fossil fuels. The United States (US) Energy Information Administration, for example, estimates that fossil fuels meet around 82% of US energy demand, and similar figures are likely to apply to other developed countries. The relative contribution of different energy sources to global energy consumption are represented in Figure 1.1. While renewable energy and nuclear power are growing rapidly, fossil fuels continue to dominate and are predicted by the US Energy Information Administration to continue to supply 80% of world energy right through to the mid-21st century. During that time, world energy use is predicted to grow 56% as gross domestic product rises, particularly in India and China, which will account for half of the predicted increase in energy use between them.

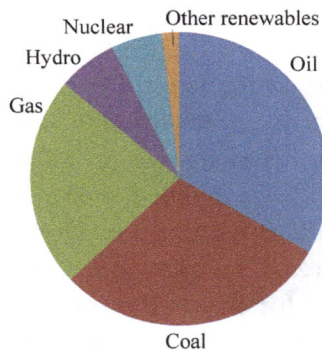

Figure 1.1   Pie-chart showing the relative contribution of different energy sources to global energy consumption in 2010 (data from US Energy Information Administration).

Coal has probably been used in China for at least 3000 years, has been a domestic fuel in Europe since the 16th century, and fuelled the industrial revolution in Europe in the 17th and 18th centuries. Today, about 7 billion tonnes of coal are produced each year, with most being used for electricity generation. Mineral oil was also used in China about three thousand years ago, mainly for light and heating, and was used for a variety of purposes in the ancient Middle Eastern civilisations. The ancient Chinese were also the first to use natural gas as a fuel, and prior to the widespread use of electricity in the 19th century, natural gas was an important domestic light and heating fuel in Europe and the US.

Drilling for petroleum oil trapped in underground reserves was first undertaken successfully in Pennsylvania in 1859 and oil soon became a popular, cheap lighting fuel. It really came into its own, however, as a source of transport fuel, with the development of the internal combustion engine. The availability of cheap oil led to the huge growth in ownership and use of private cars powered by petrol (US gasoline) and diesel, the development of diesel-powered trucks, lorries, buses, train locomotives, boats and ships; and later made international travel by air widely affordable. Current world oil production is an astonishing 13.5 billion litres (approximately 12 million tonnes) per day, with about 40% of that being used to produce petrol, 27% to produce diesel and 8% aviation fuel, with the rest being used as a raw material in manufacturing.

The industrial extraction of natural gas from underground reserves actually predates that of oil, starting in the US in 1825, but for much of the 20th century it was associated with oil extraction. Indeed, natural gas was long considered to be no more than a dangerous nuisance associated with oil drilling and was burnt off. Today it is an important domestic and industrial fuel, and global production is around 2000 billion cubic metres (1.3 billion tonnes) per year. Natural gas burns very cleanly and causes almost no pollution other than CO_2.

In recent years, trillions of cubic metres of natural gas have been discovered in shale deposits (shale is a sedimentary rock formed from compressed mud) and this has led to a huge industry springing up in North America to extract this gas using a technique called hydraulic fracturing, or fracking. In this process, a bore-hole is drilled into a shale deposit and a mixture of water, sand and chemicals, including hydrochloric acid, which dissolves minerals and initiates cracks in the shale, is pumped into the rock at very high pressure. The gas trapped in the rock is released and escapes to the head of the well, where it is collected. Fracking is controversial, but it promises to secure gas supply in the US and Canada for a hundred years and is on the way to making the US a net exporter of energy for the first time in decades. There is currently no commercial fracking in the United Kingdom (UK), but the UK has large shale deposits that almost certainly contain significant amounts of gas.

Given the huge demand for fossil fuels and the dependence on them of our modern way of life it is fortunate in terms of energy security that there are vast reserves of coal, oil and gas. Indeed, fossil fuels will have a dominant place in the energy market for decades to come. However, those reserves are not infinite and eventually they will run out. There is no definite answer as to when that will be, because it depends on energy consumption; the development of alternatives, of which biofuels are one; and the exploitation, not only of known reserves, but of projected reserves that have still to be discovered and exploited. However, problems in fossil fuel supply may occur long before reserves run out, an issue that was encapsulated in M. King Hubbert's definition of 'peak oil', the point in time when the maximum rate of petroleum extraction is reached, after which the rate of production is expected to enter terminal decline.

Hubbert was a geophysicist and he warned of impending 'peak oil' as long ago as 1956, predicting that it would be reached in the 1970s. As far as the US was concerned this turned out to be accurate, with US oil production peaking in 1970. However, the world's oil reserves have proven to be far greater than was known at the time, and peak oil has still not been reached. Nevertheless, there appears to be a consensus amongst industry experts that it will be reached in the 2020s. How painful that turns out to be will depend on actions taken now to develop alternative energy sources. Some commentators predict that energy costs will climb, affecting transport, agriculture and industry as well as domestic fuel security. More optimistic scenarios would see initial price increases being offset as alternatives are developed.

A finite supply is not the only problem associated with the burning of fossil fuels. Mineral oil, coal and gas contain carbon that was removed from the atmosphere by photosynthesising plants and microbes millions of years ago. Most carbon that is fixed in this way moves through the carbon cycle and is re-released into the atmosphere as CO_2, but some ends up in plant and animal material that is incorporated into materials such as peat, compacted into rocks and eventually mineralised to form coal, oil and natural gas. Over geological time this has reduced atmospheric CO_2 from around 7000 parts per million 540 million years ago to 270 parts per million prior to the industrial revolution.

Photosynthesis evolved when CO_2 levels were many times higher than they are today and the long-term decline in CO_2 will eventually result in its concentration falling below the point where plants can photosynthesise efficiently. There have been blips in the decline in the past, probably as a result of periods of unusual volcanic activity, such as during the so-called 'Cretaceous superplume' (a period of extreme volcanic activity involving in some definitions an upwelling of the Earth's mantle). Currently the planet is on the up-curve of another rise (Figure 1.2) because burning fossil fuels releases CO_2 back into the atmosphere, and it is the more immediate effect of this rise that is a potential threat to our way of life.

Given that plants require CO_2 to photosynthesise, a modest increase in atmospheric CO_2 levels on its own would almost certainly enhance crop productivity. However, CO_2 is a greenhouse gas, in other words it absorbs and emits radiation within the thermal infrared range. Because thermal radiation from the Earth's surface is absorbed by a greenhouse gas then re-emitted in all directions, some of it back towards the surface and lower atmosphere,

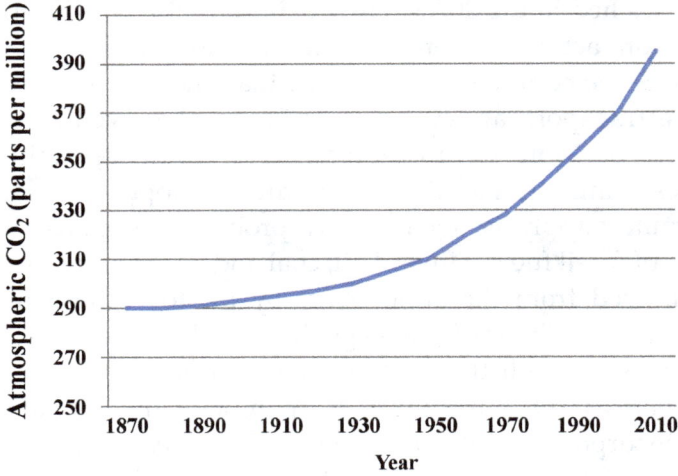

Figure 1.2   Atmospheric carbon dioxide concentration from 1870 to 2010 (data from International Panel for Climate Change (IPCC) 5th Assessment Report).

the gas causes a warming of the Earth's surface. CO_2 is not the most potent greenhouse gas (nitrous oxides and methane, for example, are many times more potent greenhouse gases), but it is the most important because of the huge volume that is being released into the atmosphere as a result of burning fossil fuels. The atmospheric CO_2 concentration has already risen from its pre-industrial level of 270 ppm (parts per million) to 390 ppm and is rising at 1–2 ppm per year, taking it to levels not seen for 20 million years.

The effect of increased CO_2 concentrations is predicted to be global warming, leading to climate change. This has been a controversial issue in recent years but that is not a topic for this book; suffice to say that it is the opinion of a substantial majority of climate scientists that there is already evidence of global warming and climate change, and that increasing CO_2 concentration as a result of human activity is the probable cause. Furthermore, if the predictions of future climate change are correct, global warming will cause changes in temperature at a rate unmatched by any temperature change over the last 50 million years. For example, the temperature changes that took place between ice ages and warm interglacial periods during the last million years were of 4 to 7 °C, but they occurred relatively gradually, with the global warming at the end of each ice age taking approximately 5000 years. In contrast, the upper end of the range predicted by Global Climate Models in the International Panel for Climate Change (IPCC) 5th Assessment Report is a 5 °C increase in global mean temperature by the end of this century. As well as this increase in temperature, there is predicted to be an increase in the frequency and severity of extreme weather events.

The potential consequences of such events for food production can be seen in the effects of the severe Australian drought of 2007–2008 and the Russian drought of 2010. In both cases, severe drought was accompanied by extreme temperature, with catastrophic effects on food production followed by peaks in the price of commodity crops. Crop prices are affected by many things, of course, including increased demand, some of which could be attributable to the use of crop products for bioenergy. This issue will come up many times in this book. However, the spikes in price caused by these two extreme weather events are clearly evident in the graph of the price of wheat shown in Figure 1.3.

1.2 Composition of Fossil Fuels

To understand the current and potential roles of biofuels, biomass and biogas in energy provision it is necessary to consider the composition and properties of the fossil fuels that they are partially replacing. The textbook description of coal is a carbonaceous, sedimentary rock, formed from accumulated vegetable matter that has been altered initially by decay and then by heat and immense pressure over many millions of years. People are often vague about its composition, perhaps because there is more than one type, but all coal is composed principally of carbon, with layers formed of inter-connected 6-carbon rings. Coal also contains varying amounts of oxygen, nitrogen and sulphur, with the nitrogen and sulphur being problematic because they give rise to pollutants during burning.

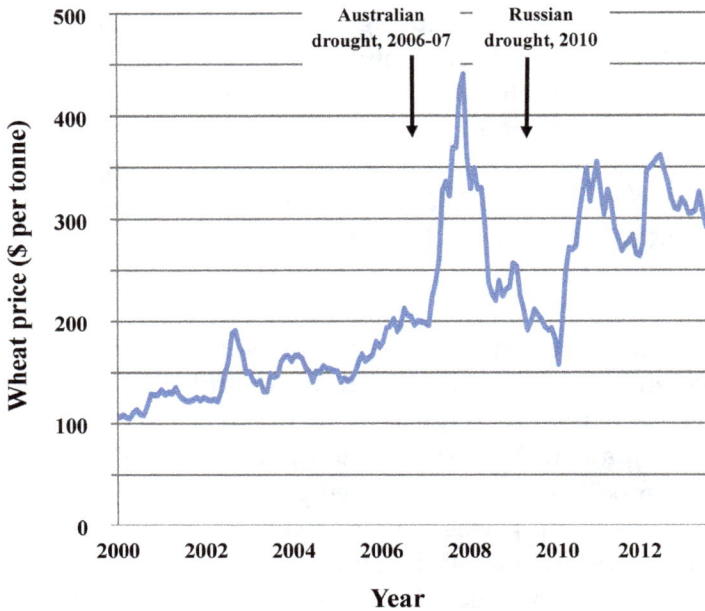

Figure 1.3   Wheat price from 2000 to 2013 (data from World Bank). The timings of extreme weather events are indicated.

The different coal types, from the lowest grade (containing more impurities) to the highest, are: lignite (brown coal), sub-bituminous coal, bituminous coal, steam coal (also known as sea coal) and anthracite. There are two other, similar forms of mineral carbon: jet and graphite. Both are composed of pure carbon and they have a number of uses, including jewelry for jet and pencil manufacture for graphite, but neither is used for fuel because they are too valuable and do not burn well anyway. Carbon will also form diamonds but by a completely different process involving intense heat and pressure deep in the Earth's mantle, after which the diamonds may reach the surface as a result of volcanic activity.

Because of its solid nature, coal is used primarily for electricity generation, usually by means of a steam turbine. The coal is burnt to heat water and convert it to steam; the pressurised steam is used to turn a rotating shaft and electricity is produced by electromagnetic induction. The largest coal-fired power station in the world is the

Taichung Power Plant in Taiwan, which can generate 5500 million watts of electricity from coal-fired steam turbines and about another 300 million watts from gas-fired and wind turbines. It consumes 14.5 million tonnes of coal per year and emits a huge 40 million tonnes of CO_2. Coal is also used as a domestic fuel for heating homes and providing hot water, and to power steam-driven locomotives (now a rarity) and ships.

Natural gas is also used as a fuel for electricity generation, and as a versatile industrial and domestic fuel that can be readily piped to and around houses and factories. Natural gas is methane, a simple hydrocarbon comprising a carbon atom and four hydrogen atoms (CH_4; Figure 1.4). Compressed natural gas (CNG) is simply methane that is compressed to 1% of its normal volume and stored

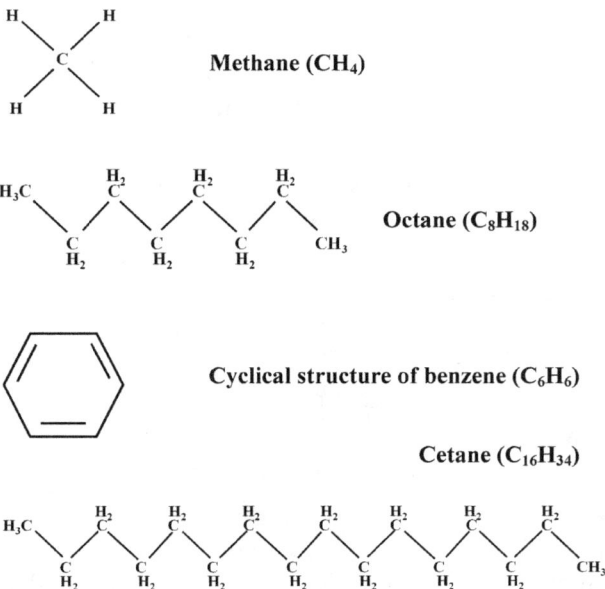

Figure 1.4 Diagrams representing the structure of methane, the longer alkanes, octane (typically present in petrol) and cetane (typically present in diesel), and the cyclic (aromatic) hydrocarbon benzene. Benzene contains three double bonds and three single bonds between the six carbon atoms making up the ring, as indicated.

at high pressure. There is increasing use of CNG as a vehicle fuel; it can replace petrol, diesel or liquid petroleum gas, producing a cleaner exhaust than any of these oil-derived fuels and is regarded as relatively safe because of its rapid dispersal if it is released as a result of a collision.

Ordinary petrol engines can be converted to use CNG, or a combination of petrol and gas, and there is increasing production of vehicles specifically designed to use it. Diesel engines can also be modified to run on diesel natural gas (DNG), a mixture of 70% CNG and 30% diesel. CNG is generally cheaper than the fuels it replaces, although this may be offset by the cost of converting the engine. The number of natural gas vehicles now exceeds 15 million worldwide, with increasing popularity in Iran, Pakistan, India, the Asia-Pacific rim, Brazil and Argentina.

Carbon can form the skeletons of myriad compounds because of its ability to form four chemical bonds. In the case of methane these are all with hydrogen atoms; but if one is with another carbon atom, and each carbon is linked to three hydrogen atoms, then a C_2H_6 hydrocarbon called ethane is formed. Addition of another carbon molecule to the chain gives propane. Since the middle carbon atom is joined to the end carbon atoms, it is bonded to only two hydrogen atoms, so the chemical formula is C_3H_8. A fourth carbon gives butane, C_4H_{10}, and so on, with the general formula for this type of molecule being C_nH_{2n+2}.

These simple hydrocarbons are called alkanes. Ethane and propane are gases at atmospheric pressure, but can be stored under pressure in liquid form. Indeed, liquefied petroleum gas is propane or a mixture of propane and butane. Longer alkanes are liquid at atmospheric pressure unless they are heated. Hydrocarbons with more carbon atoms can also form ring structures (known as cycloalkanes or aromatic hydrocarbons) or branched molecules. Crude petroleum oil contains a mixture of hydrocarbons and these are separated into fractions for different fuel types and other end uses. Petrol typically contains hydrocarbons with 4 to 12 carbon atoms. The 8-carbon molecule octane (C_8H_{18}) is shown in Figure 1.4.

Octane gives its name to a quality measure of petrol known as the octane rating. Petrol is used in internal combustion engines in which the fuel is ignited by a spark from a spark plug and burned in a controlled manner. If the fuel self-ignites due to heat and pressure, it can damage the engine. This is called engine knocking, and is reduced by using a fuel that is more resistant to self-ignition. The octane rating is a measure of this resistance and is affected by the chain length of the hydrocarbons that are present and the proportion of cyclic molecules. Note, however, that some cyclic hydrocarbons, such as benzene (C_6H_6; Figure 1.4), are carcinogenic, and their presence in petrol is restricted in many countries.

Diesel engines, like those that use petrol, are internal combustion engines, but they differ from petrol engines in that they use the heat of compression rather than a spark from a spark-plug to ignite the fuel. Diesel fuel is made up of longer-chain hydrocarbons, with 8 to 21 carbon atoms. The 16-carbon molecule cetane is shown in Figure 1.4. As with octane, cetane gives its name to a quality standard, the cetane value, which is measured using the readily flammable cetane and its much less flammable branched isomer *iso*-cetane as standards. This will be dealt with in detail in Chapter 3.

1.3 Terminology, Further Definitions and an Introduction to the Major Bioenergy Feedstocks

The term bioenergy is defined at the start of this chapter as energy produced from (recent, renewable) biological sources. Bioenergy can be subdivided in a number of ways, but in this book the terms biofuel, biomass and biogas are used. These terms are not defined consistently and there may be overlap between them anyway, but here the term biofuel is applied to liquid fuels, bioethanol and biodiesel; biomass to solid biological material used for heating and electricity generation, and biogas to a gas containing methane that is produced from the anaerobic digestion of biological, usually waste, matter.

Bioethanol is used primarily as a substitute for petrol in transport fuel. It is produced from sugars, and the major sugar crops (sugar cane and sugar beet) are important sources of feedstock. Indeed, production of ethanol from sugar cane has been an established industry in Brazil for several decades. Sugars can also be produced from starch through enzymatic digestion, which means that cereal grain and other starchy crop products represent another feedstock. Bioethanol production from maize starch is now well established in the US.

Biodiesel is derived from plant oils and waste animal fats, usually after the fatty acids present in the oils and fats have been esterified with methanol to create fatty acid methyl esters (FAMEs). Biodiesel production is now an important and rapidly growing industry in Europe and the US. All of the major oil crops are potential sources, including oilseed rape (canola), soybean and palm oil. Oil from less well-known, non-food crops may also enter this market, including *Jatropha* and *Pongamia*. *Jatropha* is the name given to a large group of succulent trees and shrubs, sometimes given the common name 'physic nut'. *Jatropha* oil is toxic but that does not affect its use for biofuel production, and *Jatropha* could be cultivated in sub-tropical and tropical countries on land that is currently considered too poor for food production. *Pongamia* is another medium-sized tree, suitable for tropical and sub-tropical cultivation, which produces large, oil-rich seeds.

Biomass is solid, mainly plant-derived material used for fuel. As such, it could be said to include traditional charcoal, which, as discussed above, is still an important fuel in many parts of the world. However, the term biomass was coined relatively recently and is more commonly applied to waste products and to novel crops, including fast-growing trees such as willow and poplar, 'giant' grasses, such as *Miscanthus* and switchgrass, and reeds. It is envisaged that these could generate material that would be used for electricity generation, replacing some of the fossil fuel that is used for that purpose and the potential market is obviously huge. So far the novel crops have not been adopted on a large scale, possibly because other 'renewable' alternatives to fossil fuels for heat and power generation, such as wind, hydro-electric and solar energy, are also being developed. However, there is an emerging market for waste material from conventional forestry to be used in electricity generation. Another distinguishing feature of biomass production is the recognition that the industry must be sustainable and not cause the deforestation associated with charcoal production.

It is worth noting here that some crops that are principally regarded at present as biofuel crops may also be used for the production of biomass and biogas. Likewise, plant material that is currently used for biomass may in future be used for the production of cellulosic ethanol (Section 2.5) or biogas, so classifications of different species as biofuel, biomass or biogas crops may change in the coming years.

Other terms with loose definitions that are applied to bioenergy crops are first and second generation. In this book, the term first generation is applied to established food crops where the crop product that would otherwise be used for food production is being used for energy. There are many examples in Chapters 2 and 3 of sugar, grain and oil crops to which this applies. In the case of bioethanol production, the term second generation is used where the feedstock is cellulose, not sugar or starch, with the feedstock coming from the inedible parts of a food crop (such as the straw) or a non-food crop. For biodiesel, the term second generation is appropriate where the feedstock is an inedible oil obtained from a novel crop grown on land that is not suitable for food production, or from algae; there are several examples of second generation biodiesel crops described in Section 3.5. The term second generation is also applicable to biomass production from sustainable, non-food sources, and biogas production from waste products or algae. Examples of first and second generation feedstocks are given in Table 1.1.

Table 1.1   **First and second generation bioenergy feedstocks.**

First Generation	Second Generation
Bioethanol	
Sugar from sugar cane and sugar beet	Cellulosic feedstocks:
Maize grain starch	Straw from cereal crops
Wheat grain starch	Bagasse and other agricultural waste
Starch from other cereal grains	Non-food crops, such as willow and
Starch from cassava storage roots	*Miscanthus*
Starch from potato and other root crops	Algae
Biodiesel	
Soybean oil	*Jatropha* oil
Oilseed rape oil	*Pongamia* oil
Palm oil	Castor bean oil
Other food oils	Algae
Biomass	
Cereal grain, such as feed wheat	Forestry waste
	Willow
	Poplar
	Miscanthus
	Other giant grasses and reeds
Biogas	
	Food waste
	Sewage gas
	Bagasse, silage and other agricultural waste
	Manure
	Algae

1.4 Potential Benefits of Replacing Fossil Fuels with Biofuel, Biomass and Biogas

The huge global demand for energy means that biofuels will only ever be able to meet a modest proportion of that demand. This is evident from the fossil fuel production figures given earlier in this chapter, and will become more clear in the following chapters as biofuel, biomass and biogas production figures are discussed. However, replacing some of the fossil fuel that is currently being used for energy production with a renewable alternative has some potential benefits, including:

- Reducing CO_2 emissions
- Reducing emissions of nitrogenous and sulphurous pollutants
- Conserving fossil fuel reserves, giving more time for alternatives to be developed
- Diversifying energy production, making energy security less dependent on any one particular industry or on fuels that are controlled by a small number of countries
- Providing additional markets for farm produce, making farming more profitable
- Enabling farmers to grow crops on 'marginal' land that would not be productive enough for food crops, again making farming more profitable

Some of these benefits are not as simple as they may first appear, starting with the first and arguably most important benefit of reducing CO_2 emissions. A basic premise of using biofuel and biomass in place of fossil fuels is that the CO_2 that is released when the biomass or fuel is burnt is first acquired from the atmosphere as the crop grows, so that, theoretically, there could be no net CO_2 emission associated with the process. In reality, most crops require a considerable input of fossil fuels for their cultivation: in the production of agrochemicals and fertilisers, the oil and diesel that is used to power farm machinery and vehicles, and in transport, storage and processing, for example. The carbon equation for different crops and production systems varies, and how it is calculated is a controversial issue, with some scientists questioning the ethics and usefulness of biofuels altogether.

Other factors may or may not be included when considering the real benefit in terms of CO_2 emissions of using a crop for energy production, including the effect of converting the land where the crop is grown from its previous use, the efficiency of the fuel that is derived compared with the fossil fuel it is replacing, and the production of co-products such as animal feed. The inclusion or exclusion of these factors can change calculations of CO_2 savings enormously. It is also important to note that benefits may increase once the industry is established and investment is made in improving the efficiency of the process and exploitation of the feedstock. The use of bioethanol from maize starch, for example, may not currently give much, if any, saving in CO_2 emissions compared with using petrol from petroleum oil, but the calculation would change completely if systems were developed to convert cellulose in the straw into bioethanol (Section 2.5).

There are several ways of describing CO_2 emissions associated with crop or energy production. In this book, the term carbon intensity (CI) (sometimes called emission intensity) is used, which is a measure of the CO_2 that is required per unit of energy that is produced, using the units g/MJ (gram per megajoule, or million joules). Some readers may be more familiar with kilowatt hours (kwh) as a unit of energy: 1 kwh = 3.6 million joules, and 1 MJ = 0.2777 kwh. CO_2 saving (or greenhouse gas saving) associated with a particular biofuel can then be calculated as:

(CI fossil fuel equivalent – CI biofuel)/CI fossil fuel equivalent

While the production of bioethanol and biodiesel from different crops is discussed in detail in subsequent chapters, the carbon intensities associated with some of the feedstocks and fuels are given in Tables 1.2 and 1.3, along with the carbon intensities of the fossil fuels that they replace. Note that these figures come from the UK's Department of Transport, and in the case of the biofuels are described as conservative estimates. They are for fuels that are burnt in the UK, so the energy use associated with transport from the country of origin to the UK is included. Note also that meaningful, independent figures are not available for the use of biomass or biogas, because those industries have not developed sufficiently to allow figures to be calculated.

One remarkable point that emerges from the analysis is that the CI of fossil fuel natural gas is only just over half that of coal, and is

Table 1.2 Conservative estimates of carbon intensity (CI) of bioethanol from different crop sources compared with fossil fuels, where the fuel is used in the UK.

Fossil fuels		
		CI (g/MJ)
	Natural gas	62
	Diesel	86
	Petrol	85
	Coal	112

Bioethanol		
Source	**Country**	**CI (g/MJ)**
Sugar beet	UK	50
	UK, British Sugar	24
Wheat	UK	61
	Ukraine	103
	Germany	59
	France	65
Sugar cane	South Africa	112
	Pakistan	115
	Mozambique	30
	Brazil	24
Maize	France	49
	US	108

Source: United Kingdom Department of Transport, apart from figure given for sugar beet by British Sugar.

lower than several of the bioethanol and biodiesel figures. Another is that the CI of a particular biofuel is very dependent on where it is produced, with the CI for bioethanol from Brazilian sugar cane, for example, being $24\,g/MJ$ but for Pakistani sugar cane being $115\,g/MJ$, compared with $85\,g/MJ$ for petrol produced from petroleum. In other words, using bioethanol from Pakistani sugar

Table 1.3.   Conservative estimates of carbon intensity (CI) of biodiesel from different crop sources compared with fossil fuels.

Fossil fuels

		CI (g/MJ)
	Natural gas	62
	Diesel	86
	Petrol	85
	Coal	112

Biodiesel		
Source	**Country**	**CI (g/MJ)**
Oilseed rape	Australia	63
	Canada	54
	Finland	52
	France	46
	Germany	47
	Poland	45
	UK	55
	Ukraine	59
Soybean	Argentina	42
	Brazil	73
	US	55
Palm	Malaysia	38
	Indonesia	38

Source: United Kingdom Department of Transport.

cane in the UK is associated with an increase, not decrease, in CO_2 emissions compared with using petroleum-derived petrol.

The tables also illustrate the lack of agreement on these figures, with the UK's Department for Transport, for example, calculating the CI for bioethanol from sugar beet at 50 g/MJ while British

Sugar calculates it for its own factory at Wissington in Norfolk as 24 g/MJ, less than half the Department for Transport figure. Similarly, the CI for bioethanol produced from US maize is hotly disputed, with the UK Department for Transport calculating a figure of 108 g/MJ, which would represent a 27% increase in CO_2 emissions compared with using petroleum-derived petrol, while a report prepared by Susan Boland and Stefan Unnasch of Life Cycle Associates for the US Renewable Fuels Association claims a 32.7% reduction in CO_2 emissions. Some of the difference would result from transportation of the ethanol to the UK; nevertheless the discrepancy is large, and the lack of consensus on CI values makes it difficult to draw firm conclusions on the impact of using biofuels on CO_2 emissions.

1.5 Political Drivers for Biofuel Development

Given the uncertainty over reductions in CO_2 emissions associated with biofuel production, the rapid growth of the biofuel industry this century, which will become evident in the following chapters, may seem surprising. However, reducing CO_2 emissions is not the only benefit associated with the use of biofuels. Indeed, the Brazilian bioethanol industry sprang up in the 1970s, long before anyone was worrying about climate change, and the US biofuel industry grew rapidly in first few years of the 21st century, to a great extent because of political support, despite the US government at the time being sceptical about the science behind warnings of climate change. The main driver for both the Brazilian and US governments appears to have been a desire to reduce their country's dependence on imported oil.

The Brazilian biofuel industry that arose in the 1970s was based on bioethanol from sugar cane (Section 2.3.2) and driven by the Brazilian Government's National Alcohol Program (known in Brazil as Proálcool) during the oil crisis of that decade. Sugar cane now provides nearly 17% of Brazil's energy. The production of bioethanol from maize starch in the US (Section 2.4) also began in the 1970s and early 1980s, again as a result of political intervention through the Energy Policy Act of 1978. This provided for a government subsidy to bioethanol in transport fuel of 40 cents per US gallon (10.5 cents per litre) and the US government has continued to subsidise the use of bioethanol in transport fuel ever since. Currently the subsidy stands at 51 cents per US gallon (13.5 cents per litre). The US government also imposes a tariff to protect the industry from competition, particularly from Brazilian ethanol produced more cheaply from sugar cane.

Bioethanol from maize starch did not take off in the 1970s in the US in the way that bioethanol from sugar cane did in Brazil, but has grown rapidly in the last decade and a half. Most recently, the Energy Independence and Security Act of 2007 established a renewable fuel standard calling for a quadrupling in bioethanol production to 36 billion US gallons (136 billion litres) annually by 2022 (the so-called Ethanol Mandate). This Act also showed that the US government was becoming more sensitive to the issue of climate change because, for the first time, it required that renewable fuels produced in facilities built after the date of enactment achieve at least a 20% reduction in greenhouse gas emissions, including indirect emissions associated with, for example, land use change.

The European Union has also adopted a policy of supporting the increased use of renewable energy sources and created an artificial market for biofuels under Directive 2003/30/EC, which established the goal of reaching a 5.75% share of renewable energy in the transport sector by 2010. A second directive, Directive 2009/28/EC on the promotion of the use of energy from renewable sources, set the goal of a share of a minimum of 10% for every Member State to achieve by 2020. This second directive stipulated the use of biofuels that generate a clear greenhouse gas saving without affecting biodiversity and land use.

The UK government responded to the EU's renewable energy directives by establishing Renewables Obligation Schemes. For the transport sector, the scheme is called The Renewable Transport Fuel Obligation and it imposes a legal obligation on road transport fuel suppliers to produce Renewable Transport Fuel Certificates demonstrating that renewable fuel has been supplied up to a specified percentage of their total fuel sales. The intention is that the specified percentage will be increased over time to meet the targets set in the EU directives. If suppliers do not achieve the specified percentage, they have to buy out of their obligation at a price per certificate set by the Office of the Renewable Fuels Agency.

Without these political interventions, biofuels for the transport sector would have to be sold into a highly competitive market against the background of enormous volatility in the price of petroleum oil, and it is highly unlikely that the biofuel industry would have attracted the necessary investment to develop in the way that it has. There has also been a political push to promote the use of biomass for electricity generation. In the UK, as with transport fuel, there are Renewables Obligation Schemes for this sector, requiring electricity suppliers to generate a proportion of their electricity from renewable sources. In England, Wales and Scotland this proportion was set at 3% in 2002 and will reach 15.4% by 2015 or 2016, while the proportion in Northern Ireland will be 6.3%. The schemes are considered to have been a success, increasing the level of renewables used in electricity generation from 1.8% in 2002 to 7% by 2010, and have been extended to run until 2037 through the UK government's Renewable Energy Action Plan.

The energy sources that qualify as renewables in this sector include biomass and biogas, but also hydroelectric, tidal, wind and solar energy. The schemes are operated by the Office of Gas and Electricity Markets (Ofgem), which issues a Renewables Obligation Certificate for every megawatt-hour of electricity produced from renewable sources (although some sources, such as offshore wind, qualify for more, while others, such as sewage gas, qualify for less). Companies that fail to meet their obligation can buy out their unfulfilled certificates, currently at a cost of £42.02 per certificate. The UK also operates a Feed-in Tariffs scheme, which provides a guaranteed price for a 20-year term for small-scale electricity generators using renewable feedstocks, and Renewable Heat Incentives, which provide a fixed income to generators of renewable heat.

In the US, the United States Department for Agriculture's Biomass Crop Assistance Program (BCAP) provides federal funding incentives to farmers to grow non-food crops from biomass.

1.6 Food versus Fuel

As discussed at the start of this chapter, the use of plant material for fuel is nothing new. However, the dependence on plant products for fuel, at least in developed countries, declined as alternative sources of energy were discovered. As a result, fuel and food production from agriculture have not generally been in conflict before. That is not the case now, with agriculture expected to meet the challenge of providing food security for an expanding world population while also helping to meet the demand for fuel. Competition for land, water, fertiliser and other inputs also exacerbates the problem.

Bioenergy production was originally envisaged as being centred on dedicated non-food crops with the potential to produce high biomass yields with relatively low fertiliser inputs, such as willow, poplar and some perennial grasses (Chapter 4). However, it is the use of sugar cane and starchy grain crops such as wheat and maize to produce bioethanol and oil crops for biodiesel that has taken off first. These crops are, obviously, also used for food. Furthermore, the global biofuel industry has expanded rapidly since the turn of the century, during a period that has also seen a steady rise in food consumption. Figure 1.5, for example, shows the consumption of wheat from 1960 to 2013, during which time global consumption more than trebled. World population more than doubled during this period, from 3.026 billion to 7.162 billion (data from UN Department of Economic and Social Affairs, Population Division) and *per capita* consumption increased by 34%, from 73 to 98 kg per year. The increasing demand has added to the effect of severe weather events discussed earlier in this chapter in pushing prices up (Figure 1.3).

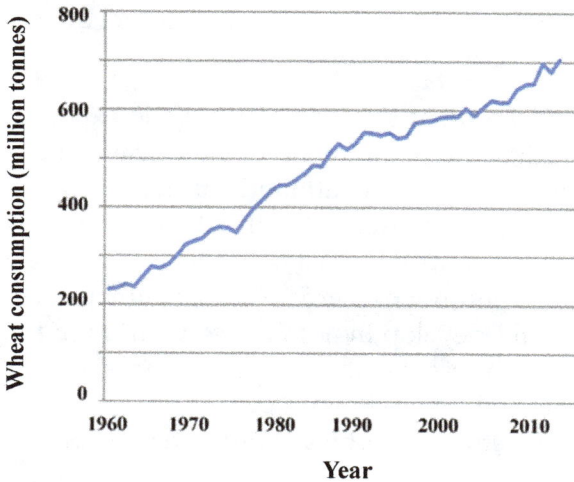

Figure 1.5 Global wheat consumption from 1960 to 2013 (Data from United States Department of Agriculture).

Biofuel production has had very little direct influence on wheat consumption to date. That is likely to change as increasing amounts of wheat grain are used to make bioethanol (Chapter 2.4.5), but the increase in consumption shown in Figure 1.5 has arisen from population growth and increased prosperity, as countries such as China and India, with huge populations, develop economically and people have more money to spend on food. People with more spending power not only want to eat more but also want to eat better and in particular to eat more meat: it takes much more grain to rear an animal for meat than it does to feed people directly, but meat provides high quality protein, vitamins, iron and other minerals, and is valued for its flavour and texture.

Nevertheless, the use of some food crops for biofuel production has contributed to increasing demand and the food versus fuel issue has become a contentious one that is discussed many times in the following chapters. It led the Nuffield Council on Bioethics, London, to publish a detailed report on 'Biofuels: ethical issues' in

2011,[1] in which an ethical framework for biofuel development was set out, stating that:

- Biofuels development should not be at the expense of people's essential rights (including access to sufficient food and water, health rights, work rights and land entitlements)
- Biofuels should be environmentally sustainable
- Biofuels should contribute to a net reduction of total greenhouse gas emissions and not exacerbate global climate change
- Biofuels should develop in accordance with trade principles that are fair and recognise the rights of people to just reward (including labour rights and intellectual property rights)
- Costs and benefits of biofuels should be distributed in an equitable way

The ethical issues associated with biofuel and more generally bioenergy production will be discussed further in the following chapters, which describe in detail the development of the global bioethanol, biodiesel, biomass and biogas industries.

---

[1] Nuffield Council on Bioethics (2011). *Biofuels: Ethical Issues*. Nuffield Press, Abingdon, Oxfordshire, UK.

2 BIOETHANOL

2.1 Introduction

Ethanol (ethyl alcohol) (Figure 2.1) is of course familiar as the alcohol found in alcoholic drinks. Bioethanol is just the term that has been applied to ethanol derived from crop sources and used for fuel, distinguishing it from ethanol derived from petroleum oil or ethanol in drinks. In other words, bioethanol and ethanol are chemically identical. The major use of bioethanol is as a vehicle fuel, usually blended with petrol. The proportion of ethanol in the mixture is 5–10% in most countries: higher ratios could cause problems with conventional petrol engines because the energy value of ethanol is less than two thirds that of petrol (19.6 MJ per litre compared with 32 MJ per litre). However, it has been mandatory to mix ethanol into petrol-based vehicle fuels in Brazil since 1976 and the current legal minimum ratio is 25% ethanol, 75% petrol.

The drive to use home-produced bioethanol in Brazil has also led to the development of so-called flex-fuel vehicles. The market for these vehicles is expanding rapidly, not only in Brazil but also in the US, Canada and to a lesser extent Europe, with Sweden leading the European countries. Flex-fuel vehicles have modified engines that can run on pure or nearly pure ethanol, and a fuel comprising 100% ethanol (E100) is sold in Brazil. In the US, Canada and Sweden the preferred blend is 85% ethanol, 15% petrol (E85) because this increases the temperature at which the fuel burns, reducing emissions and

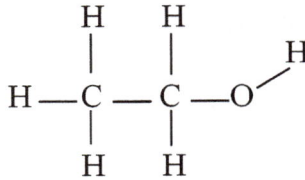

Figure 2.1   Simple representation of the chemical structure of ethanol.

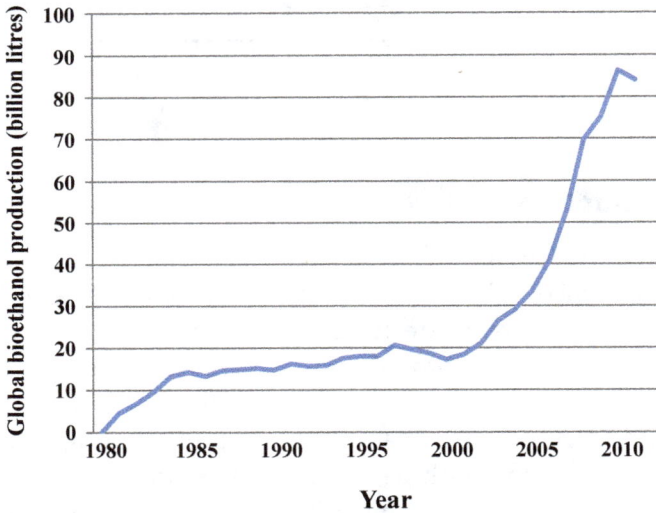

Figure 2.2   Global annual bioethanol production, 1980–2011 (data from United States Energy Information Administration).

improving starting in cold weather. Both the US and Sweden switch to lower ethanol blends in the winter (E70 and E75, respectively) because of the problems with cold-starting with high ethanol blends.

World bioethanol production has soared since the turn of the century, from approximately 17 billion litres in 2000 to over 84 billion litres in 2011 (Figure 2.2). The US is by far the largest producer, making over 52 billion litres in 2011, with Brazil second on over 22 billion litres, representing 63% and 27% of the total, respectively (Table 2.1). These figures may seem huge, but bioethanol production is still dwarfed by global petrol production and consumption, and bioethanol currently makes up only 7% of petrol-type transport fuel globally and about 10% in the US.

Table 2.1   World bioethanol production 2011.

Country/area	Production (billion litres per year)
US	52.73
Brazil	22.75
China	2.26
Canada	1.74
France	1.01
Germany	0.77
Australia	0.44
Belgium	0.38
Sweden	0.20
Poland	0.17
Argentina	0.12
Other	1.49
Total	84.06

Source: Food and Agriculture Organisation of the United Nations

2.2 Fermentation of Sugars to Ethanol

Bioethanol is produced from glucose and fructose through fermentation by yeast; essentially the same process that has been used for thousands of years to make alcoholic beverages. If the feedstock is sucrose (Figure 2.3), a disaccharide of glucose and fructose units, the yeast first hydrolyses the sucrose to form glucose and fructose through the action of the enzyme invertase. This reaction is irreversible and the name invertase derives from the mixture of glucose and fructose produced by the reaction, which in the food industry is referred to as inverted sugar syrup. The glucose and fructose are substrates for hexokinases, which catalyse the phosphorylation of hexoses to produce hexose phosphates to supply glycolysis, the glucose 6-phosphate that is produced from glucose first being converted to fructose

Sucrose

Glucose

Fructose

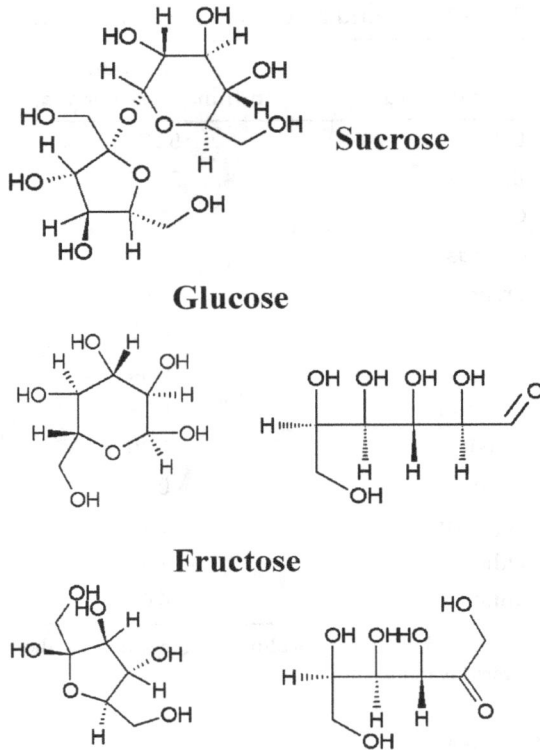

Figure 2.3   Structure of sucrose, a disaccharide of glucose and fructose units, and of glucose and fructose themselves. Glucose and fructose are shown in ring and open chain forms, in the D- (as opposed to L-) configuration, which is the configuration that occurs in nature. D-glucose is also known as dextrose.

6-phosphate by glucose 6-phosphate isomerase. In glycolysis, fructose 6-phosphate is converted through multiple steps to the three-carbon molecule, pyruvate. Overall, two molecules of adenosine triphosphate (ATP), the energy currency of eukaryotic cells, are produced, the net reaction for glucose being:

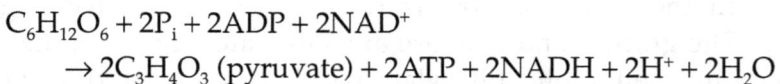

$$C_6H_{12}O_6 + 2P_i + 2ADP + 2NAD^+$$
$$\rightarrow 2C_3H_4O_3 \text{ (pyruvate)} + 2ATP + 2NADH + 2H^+ + 2H_2O$$

Note that the equation appears unbalanced, but the two phosphate (Pi) groups exist as hydrogen phosphate anions (HPO_4^{2-}).

Transfer of the phosphate group to adenosine diphosphate (ADP) gives one H^+ ion and frees an oxygen atom. This balances the equation.

In most eukaryotes under aerobic conditions, pyruvate enters the mitochondria and undergoes oxidative carboxylation to make the two-carbon molecule, acetyl-Coenzyme A (acetyl-CoA), which then enters the citric acid cycle (also known as the Krebs cycle or the tricarboxylic acid cycle) and is completely oxidised to CO_2 and H_2O. It is here that most of the energy in the sugar molecule is released. However, if yeast is adequately nourished it converts the pyruvate to a different two-carbon molecule, acetaldehyde, a reaction catalysed by pyruvate decarboxylase. The acetaldehyde is then converted to ethanol by alcohol dehydrogenase. The overall conversion of glucose or fructose to ethanol is called alcoholic fermentation, and the net reaction is:

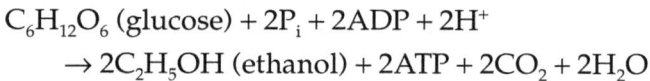

$$C_6H_{12}O_6 \text{ (glucose)} + 2P_i + 2ADP + 2H^+$$
$$\rightarrow 2C_2H_5OH \text{ (ethanol)} + 2ATP + 2CO_2 + 2H_2O$$

The whole process only generates two molecules of ATP, so most of the energy that was available in the sugar is still present in the ethanol, hence ethanol's usefulness as a fuel. Combustion of ethanol produces CO_2 and water vapour:

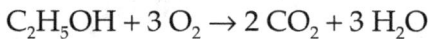

$$C_2H_5OH + 3\,O_2 \rightarrow 2\,CO_2 + 3\,H_2O$$

Yeast dies if ethanol accumulates in the fermenter beyond a concentration of approximately 15%. Ethanol is therefore constantly removed together with water and distilled to produce a 95–96% ethanol/water mixture. Distillation will not produce 100% ethanol because of the formation of a water–ethanol azeotrope (a mixture of liquids that cannot be separated by distillation), known as hydrous ethanol, which has the same boiling point as ethanol itself. Hydrous ethanol will combust but does not mix with petrol as well as pure ethanol does. Another method is therefore used to dehydrate the ethanol further.

Dehydration can be achieved by a process known as azeotropic distillation, in which benzene (Figure 1.4) is added, leading to the formation of a mixture from which pure, anhydrous ethanol can be extracted by distillation. Another method, called extractive distillation, involves adding a solvent such as ethylene glycol, which has a relatively high boiling point and interacts differently with the ethanol and water, causing their relative volatilities to change and allowing the ethanol to be distilled off. The current method of choice, however, is to use a molecular sieve: that is, to heat the mixture and pass the vapour under pressure through a bed of beads with pores that allow absorption of water while excluding ethanol. The anhydrous ethanol vapour is then condensed. The bed of beads can be reactivated by removing the absorbed water under vacuum, and two beds are often used so that one can be regenerating while the other is in use. The material used to make the beads is a water-absorbing mineral or zeolite. The required pore size for dehydrating ethanol is 3 Å (3×10^{-10} m), and this is produced from a mixture of aluminium, potassium, sodium and silicon oxides.

2.3 Bioethanol from Sucrose-Accumulating Crops

2.3.1 *Sucrose synthesis in plants*

The most direct way of producing bioethanol from a crop-derived raw material is from sucrose. Sucrose, a disaccharide comprising glucose and fructose units (Figure 2.3), is by far the most abundant sugar in plants (this characteristic is unique to plants: other organisms do not synthesise or accumulate sucrose), and the major transport molecule for moving carbon fixed in the leaves by photosynthesis to carbon 'sink' organs such as seeds, tubers, flowers and roots. Sucrose is particularly suited to this role because its solutions have relatively low viscosity. It is also relatively unreactive compared with glucose and fructose, which have highly reactive carbonyl (C=O) groups (Figure 2.3).

The pathway for sucrose synthesis is shown in Figure 2.4. While almost all plant tissues have the capability of synthesising sucrose, most net synthesis occurs in photosynthetic leaves. The triose phosphate, dihydroxyacteone phosphate (DHAP), produced in the Calvin

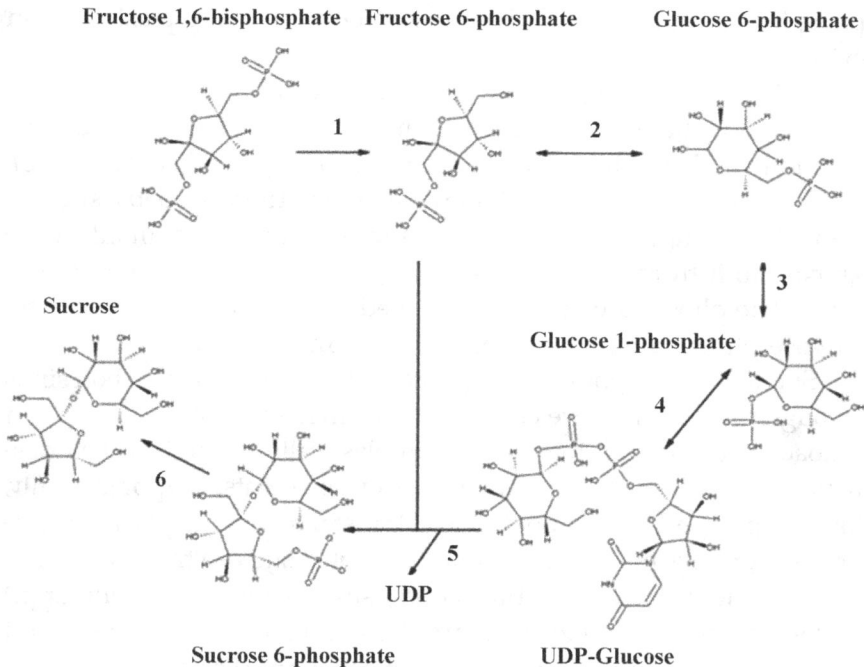

Figure 2.4   **Pathway for sucrose synthesis in plants. The enzymes responsible for the different steps are: 1. Fructose 1,6-bisphosphatase (FBPase); 2. Glucose 6-phosphate isomerase (phosphoglucose isomerase; PGI); 3. phosphoglucomutase (PGM); 4. UDP-glucose pyrophosphorylase; 5. Sucrose phosphate synthase (SPS); 6. Sucrose phosphate phosphatase (SPP).**

cycle, is exported from the chloroplasts and equilibrates with another triose phosphate, glyceraldehyde 3-phosphate (GAP) as a result of the action of triose phosphate isomerase. DHAP and GAP combine in an aldol condensation to produce fructose 1,6-bisphosphate. This is followed by the conversion of fructose 1,6-bisphosphate to fructose 6-phosphate (F 6-P), a reaction catalysed by cytosolic fructose 1,6-bisphosphatase (FBPase). F 6-P can be reversibly converted to glucose 6-phosphate (G 6-P), the interconversion being catalysed by glucose 6-phosphate isomerase (also known as phosphoglucose isomerase, or PGI, a dimeric enzyme with plastidic and cytosolic isoforms). G 6-P also interconverts with glucose 1-phosphate (G 1-P), facilitated by phosphoglucomutase (PGM), another enzyme with plastidic and cytosolic forms.

G 1-P reacts with uridine triphosphate (UTP) to produce uridine diphosphate (UDP)-glucose and pyrophosphate, a reversible reaction catalysed by UDP-glucose pyrophosphorylase (UGPase). UDP-glucose and fructose 6-phosphate are then used by sucrose phosphate synthase (SPS) to make sucrose phosphate; in effect the glucose unit from UDP-glucose being transferred to the F 6-P molecule. The phosphate group is removed from sucrose phosphate by sucrose phosphate phosphatase (SPP) to make sucrose.

Sucrose is transported from photosynthetic tissues to carbon sinks through the phloem sieve elements. Phloem loading at the source and unloading at the sink may occur symplastically through plasmodesmata linking mesophyll cells with the sieve elements, or apoplastically, involving export to the apoplast and transfer across the plasma membrane, a process mediated by proton-coupled sucrose transporters.

The sucrose-accumulating crops, sugar cane (*Saccharum* spp.) and sugar beet (*Beta vulgaris*), are the main source of sucrose, and there is potential for the use of sweet sorghum (*Sorghum bicolor*).

2.3.2 *Sugar cane*

Sugar cane (*Saccharum* spp.) is actually a group of at least six species (some taxonomists list 37) of perennial grasses (*Poaceae*) belonging to the genus *Saccharum* and tribe *Andropogoneae*. It is generally tall, up to 6 metres in height, and has thick, fibrous stalks in which sucrose accumulates. Sugar cane originated in Asia but is now cultivated all over the tropics and accounts for 80% of global sugar production. Indeed, the United Nations Food and Agriculture Organisation estimated that it was cultivated on just under 24 million hectares in more than 90 countries in 2010, with global production at approximately 1.7 billion tonnes. Brazil is the largest producer (approximately 550 million tonnes), followed by India (355 million tonnes) and China (114 million tonnes).

The different sugar cane species interbreed readily and most modern commercial varieties are hybrids. In Brazil, the most

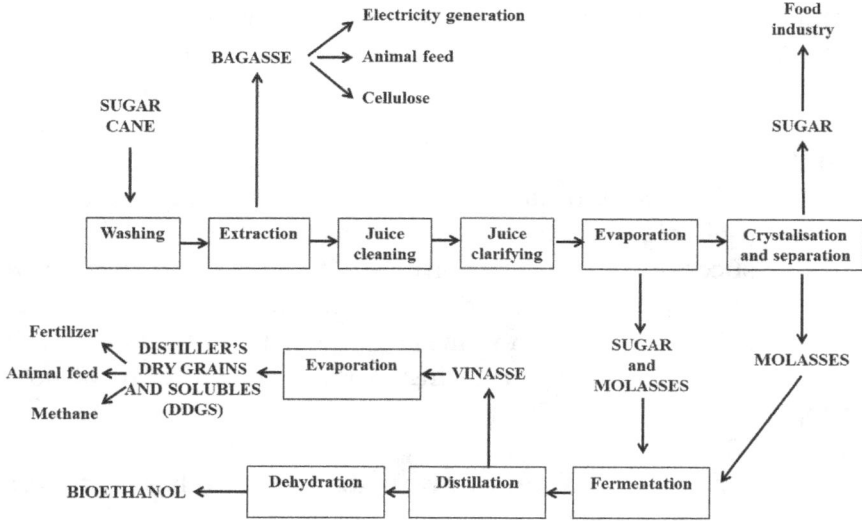

Figure 2.5   Flow-chart for the production of bioethanol and co-products from sugar cane.

widely grown hybrids result from crosses between *Saccharum officinarum* and *Saccharum spontaneum*. These are popular because they have high productivity and sucrose content, the two traits that have long been the most valued by farmers and are the main targets for breeders.

A flow-chart of the sugar production process is shown in Figure 2.5. The cane stalks are crushed and sugar-rich juice is extracted, cleaned, clarified and concentrated by evaporation. In dedicated bioethanol production plants, the sugar and molasses (a viscous by-product) are fermented to produce the bioethanol, but high-grade sugar may be crystallised out first and used for food, while the molasses and low-sugar concentration juices are used to feed the fermenter. The economic viability of the process and the carbon intensity (Section 1.4) depend on uses being found for all of the co-products. The bagasse (residual pulp) may be used for animal feed or as biomass to generate electricity (Chapter 4), often to power the plant itself, and in the future may be a source of cellulose that can be broken down into sugars and fed into the fermenter (see Section 2.5). These potential uses are leading to some breeding programmes being redirected to produce fibre-rich, so-called 'energy cane'. The molasses are fermented and the vinasse (slurry left after the distillation process) is used to make distillers dried grains and solubles (DDGS), which are used for animal feed, fertiliser or increasingly for further energy production via anaerobic digestion to produce biogas (Chapter 5).

The success of the Proálcool programme resulted in an 80% increase in sugar cane production in Brazil from 68 million tonnes in 1975–1976 to 124 million tonnes five years later in 1980–1981. It also set off a process of increasing productivity in terms of both tonnage per hectare, which has increased from 46 tonnes per hectare in 1976 to more than 75 tonnes per hectare today, and litre of ethanol per tonne of harvested raw material, which has risen from 58 litres in 1976 to over 80 litres today. Increased crop productivity has been achieved through a combination of varietal improvement and more intensive agriculture, involving the use of machinery, fertilisers, herbicides and pesticides, while improved fuel production per tonne of raw material has been achieved through the use of more efficient micro-organisms and innovations in chemical engineering. As a result, the volume of ethanol produced per hectare has more than doubled, from less than 3000 litres to more than 6000 and, according to the Brazilian Ministry of Mines and Energy, sugar cane now provides nearly 17% of Brazil's energy.

These figures show Brazil to be a world leader in generating energy from a renewable crop source and there is considerable scope for further development. The land that has been taken into sugar cane production to meet the objectives of the Proálcool programme was largely low intensity pasture and land already being used to cultivate other crops. The impact on biodiversity of converting such land to intensive agriculture is, however, a controversial, complex and divisive issue, as is the net benefit in terms of greenhouse gas emissions.

Estimates of the carbon intensity for bioethanol production from sugar cane range widely. Agricultural inputs such as fertilisers, herbicides, pesticides, machinery, fuel for agrochemical production and application, harvesting and transportation, and the energy requirement of the fermentation plant have to be taken into account. Nevertheless, the carbon intensity for ethanol production from sugar cane in Brazil, estimated conservatively by the UK's Department for Transport, is impressive at 24 g CO_2/MJ (Table 1.2); particularly considering that this figure is for the bioethanol used in the UK and so includes transport of the fuel. This gives a greenhouse gas saving of 72% when replacing petrol. This saving would be even higher if the cellulose in the bagasse were used to produce additional ethanol and the straw used as biomass for electricity generation or biogas, as is beginning to happen. The Brazilian Agricultural Research Corporation (EMBRAPA) is also developing inoculants of diazotrophic (nitrogen-fixing) bacteria to fix atmospheric nitrogen in the soil and reduce the consumption of nitrogen fertiliser, one of the largest contributors to fossil fuel consumption in sugar cane production.

Ethanol production from sugar cane in Brazil does seem to tick the right boxes in terms of land use, energy value compared with energy inputs and carbon intensity. However, while Mozambique achieves a similar carbon intensity figure of 30 g/MJ, the figures for South Africa and Pakistan are less impressive, at 112 and 115 g/MJ, respectively. Indeed, the carbon intensities for South Africa and Pakistan are higher than that for petrol (85 g CO_2/MJ), meaning that more greenhouse gases are produced by substituting petrol with bioethanol from these countries in the UK, not less.

2.3.3 *Sugar beet*

The other major source of sugar is sugar beet (*Beta vulgaris*). Sugar beet is a member of the *Amaranthaceae* family, sub-family *Chenopodiaceae*, and unlike sugar cane is a temperate crop. It is therefore the major sugar crop grown in temperate regions of Europe and North America. Sucrose accumulates in a large storage root and, unlike in sugar cane, can be extracted in a pure form that does not require further refining. *Beta vulgaris* was first grown as a fodder crop (sugar beet and fodder beet are variants of the same species), with sugary varieties not being bred until the 18th century. At that time, sucrose represented 4% of the fresh weight of the storage root, but by the mid-19th century breeding had resulted in this figure rising to 20%. Surprisingly it has not been improved further in the subsequent century and a half, with breeding focussing on yield and disease resistance. Extraction of the sucrose leaves a pulp (bagasse) that is used as an animal feed and, as with sugar cane, this is potentially an important co-product. However, beet pulp has to compete with co-products from other crops, such as wheat, maize and oilseed rape, for the feed market. An alternative may be to use the pulp for biogas production (Chapter 5).

Typical yield of sugar beet in the UK is between 50 and 100 tonnes per hectare, from which 8 to 18 tonnes of sugar are extracted. Cold periods (< 12°C) cause beet to switch to its reproductive phase and bolt, reducing yield. Yield may therefore be much higher in warmer countries with longer growing seasons. In Chile, for example, the average yield in 2010 was over 87 tonnes per hectare. The UN's Food and Agriculture Organisation lists Russia, France, the US, Germany, and the Ukraine as the world's five largest sugar beet producers, with global production in 2011 of 271 million tonnes.

In Europe, sugar beet is regarded as a good break crop in crop rotations dominated by cereal production. It is more tolerant of drought than most crops, making it suitable for soils with poor water retention, and has a relatively low requirement for nitrogen, which is important when the carbon intensity figure is calculated. On the other hand, the short season in northern Europe coupled with the expense of long-term storage and transport make bioethanol production uneconomic except as a co-product in the sugar production process. Hence, bioethanol production from sugar beet is currently closely tied in with sugar production for food use in a similar way to that shown for sugar cane in Figure 2.5. The roots are washed, sliced and boiled to extract most of the sugar, which is then crystallised, precipitated by centrifugation and packaged for supply to the food industry. The remaining liquid molasses still contain useful amounts of sugar, and this is extracted by repeating the process once or twice more. This produces relatively low concentration sugar solutions that are used to supply the fermenters that produce bioethanol.

Interest in using sugar from sugar beet for bioethanol production has been stimulated in part by the tight political control of the sugar market, which limits the opportunities for sugar trade and export. Sugar beet production in the European Union, for example, is limited to 13.3 million tonnes per year and controlled by a quota system applied to each Member State, with prices set in Brussels. In the US, 'excess' sugar production is avoided by paying farmers to destroy some of their crop if there is a good harvest. Such political interference may become less relevant as commodity prices rise. The world market price for white sugar passed the European Union sugar intervention price of €404.4 per tonne in 2009, although it has since fallen back to around €383 per tonne.

The UK's Department for Transport calculates a carbon intensity figure for bioethanol from sugar beet in the UK of $50\,g\,CO_2/MJ$ (Table 1.2), giving a greenhouse gas saving of 41% when the ethanol is used to replace petrol, although British Sugar claims a lower carbon intensity figure of $24\,g\,CO_2/MJ$ for its Wissington factory in Norfolk, UK, which is competitive with Brazilian sugar cane and gives a greenhouse gas saving of 72%.

2.3.4 *Sweet sorghum*

Another potential source of sucrose for bioethanol production is sweet sorghum (*Sorghum bicolor*). Sorghum, like sugar cane, is a grass (family *Poaceae*) belonging to the tribe *Andropogoneae*. It is closely related to maize, and may be more familiar as a grain and forage crop. Sorghum was first domesticated in what is now Ethiopia and the Sudan, and its grain is an important human food source, particularly in arid regions of Africa, Asia and Central America, where it is preferred to maize because of its better drought tolerance. This ability to withstand drought comes from its extensive root system, waxy leaves and stems and narrow leaf shape. However, it also thrives in humid conditions.

Sweet and grain sorghum varieties are variants of the same species. Nowadays most cultivated grain sorghum varieties are hybrids while sweet sorghum varieties generally are not, although there has recently been progress in developing hybrid sweet sorghum in the US. Sweet and grain sorghum are differentiated by the amount of sugar in their stalks, which is higher in sweet sorghum, and grain yield, which is higher in grain sorghum. Sweet sorghum varieties also tend to be taller because grain sorghum has been selected for reduced plant height to facilitate harvesting and reduce lodging (in which the stalk falls over in the wind and rots on the ground).

Sorghum is grown extensively in the southern states of the US, but has been cultivated as far north as southern Canada. Traditionally it has provided forage and hay, with the grain used for animal feed, but, more recently, increasing amounts of sorghum grain have been used to make bioethanol from starch (Section 2.4). Sweet sorghum was introduced from Europe in the mid-19th century in order to extend sugar production beyond the northerly limit of sugar cane cultivation. Sugar-rich juice is extracted from sorghum, as from sugar cane, by crushing and squeezing the stalks using roller mills, screw presses or diffusers, which pump hot water over shredded cane moving on a conveyor belt. The juice is then refined by evaporating off some of the water to produce syrup. Syrup is the preferred product because the stems contain high concentrations of glucose and fructose as well as sucrose, due to invertase activity, and this prevents crystallisation of the sucrose to produce granulated sugar.

There are research programmes aimed at developing sweet sorghum for bioethanol production in the US, India, Brazil and China, but currently, commercial use is small. The short harvest period poses a problem in supplying fermenters all year round, as it does with sugar beet, although the harvest period can be extended to some extent by planting multiple cultivars that mature at different rates, and by staggering planting times. This problem is exacerbated by the logistics and expense of transporting and storing the relatively bulky juice and syrup. Fermenting the sucrose in sweet sorghum juice/syrup is simpler than processing grain starch (Section 2.4), but currently maize grain is economically the better option for farmers supplying the bioethanol industry. On the other hand, while meaningful figures are not available yet, the high productivity of sweet sorghum and its relatively low nitrogen demand mean that the carbon intensity for bioethanol production from sweet sorghum is expected to be much lower than that of maize grain.

As with sugar cane, the carbon intensity and economic viability of the process depend greatly on finding uses for all of the co-products. As with sugar cane, these co-products are likely to be used as animal feed, to supply biomass for electricity production (Chapter 4), and to be composted to make fertiliser, with long-term possibilities for biogas production from anaerobic digestion (Chapter 5) and ethanol production from cellulose (Section 2.5).

2.4 Bioethanol from Starch

2.4.1 *Starch synthesis in plants*

Starch (Figure 2.6) is the major storage polysaccharide of cereal grain, potato and sweet potato tubers, cassava storage roots, a variety of other plant storage organs, and even pollen. The pathway for its synthesis is shown in Figure 2.7, but bear in mind that part of the process occurs in a specialised plastid called the amyloplast. Starch is a complex polysaccharide comprising linear chains of glucose units joined by α-1,4-glycosidic bonds (amylose) and branched molecules consisting of short linear chains linked at branch-points by α-1,6-glycosidic bonds (amylopectin) (Figure 2.6). The numbers refer to the position of the carbon atoms in each glucose unit that participate in the bond (refer to the representation of glucose in Figure 2.6). The building blocks for starch synthesis arrive from the photosynthesising leaves in the form of sucrose. The enzyme, invertase, which cleaves sucrose to glucose and fructose, has already been discussed in this chapter; but plants possess a second enzyme that cleaves sucrose: sucrose synthase (SuSy). SuSy is present in the cell wall and cytoplasm and catalyses the

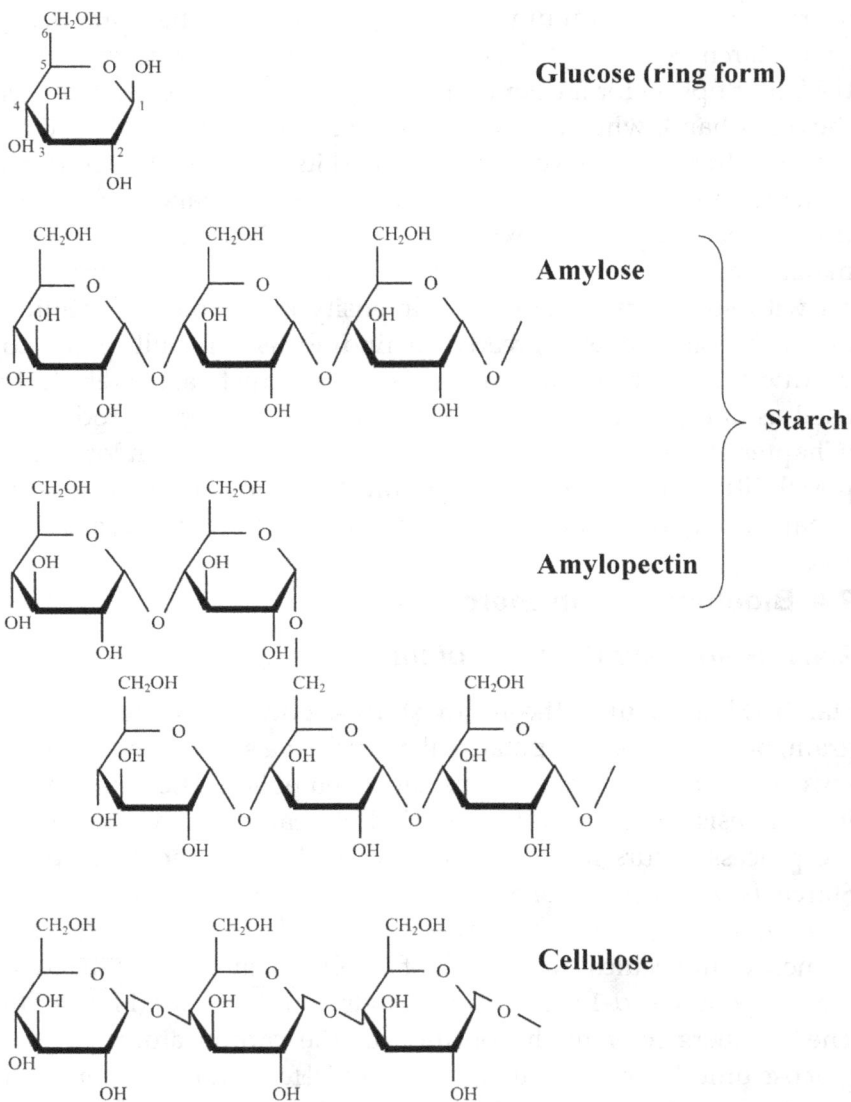

Figure 2.6   **Diagrams showing the structure of glucose in its ring form, starch (amylose and amylopectin) and cellulose. The numbers shown on the glucose molecule indicate the position of the carbon atoms. Note that individual hydrogen atoms bonded to the carbon atoms are not shown.**

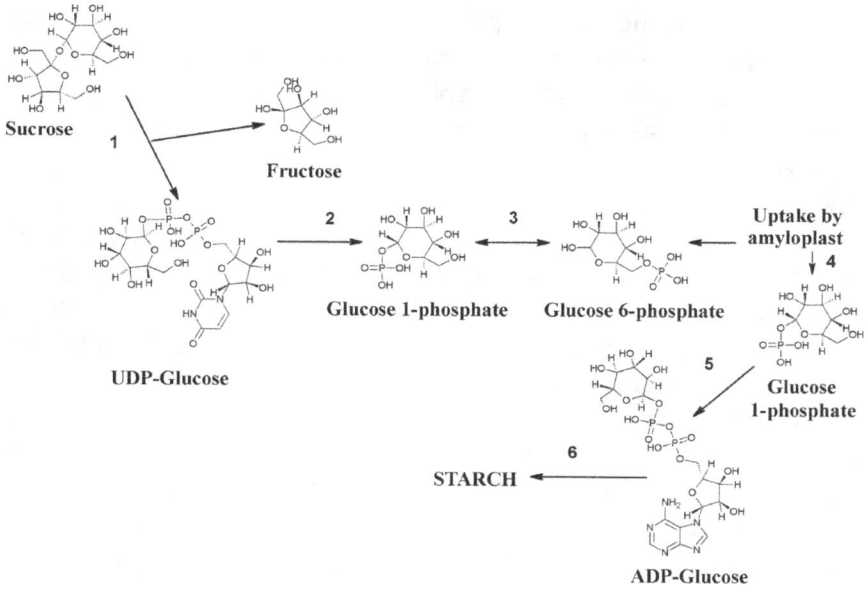

Figure 2.7   The pathway for starch biosynthesis. The enzymes involved are: 1. Sucrose synthase (SuSy); 2. UDP-glucose pyrophosphorylase; 3. Cytosolic phosphoglucomutase (PGM); 4. Plastidic phosphoglucomutase; 5. ADP-glucose pyrophosphorylase (AGPase); 6. Granule-bound and soluble starch synthases. Note that cereal grain contains a cytosolic ADP-glucose pyrophosphorylase, so steps 4 and 5 are not required and ADP-glucose is taken up by the amyloplast.

reversible conversion of sucrose and uridine diphosphate (UDP) to UDP-glucose and fructose. Despite the name of the enzyme, the equilibrium of the reaction under physiological conditions very much favours cleavage over synthesis, particularly in storage and vascular tissues where sucrose concentrations are high. The production of UDP-glucose from sucrose is the first step in the predominant route for starch synthesis; indeed, SuSy activity is closely correlated with starch accumulation.

The next step in the process is the conversion of UDP-glucose to glucose 1-phosphate, the reversible conversion that is catalysed by UDP-glucose pyrophosphorylase. Glucose 1-phosphate is the molecule that, together with adenosine triphosphate, is used by another enzyme, adenosine diphosphate (ADP)-glucose pyrophosphorylase, to make ADP-glucose and pyrophosphate. In most plant storage organs, this reaction occurs in a plastid called the amyloplast, and it appears that another sugar phosphate, glucose 6-phosphate, is taken up by amyloplasts in preference to glucose 1-phosphate. So, glucose 1-phosphate is first converted to glucose 6-phosphate by the enzyme phosphoglucomutase and the glucose 6-phosphate is converted back to glucose 1-phosphate by a similar enzyme once it has entered the amyloplast. Cereal grain, on the other hand, contains a cytosolic ADP-glucose pyrophosphorylase, so the glucose 1-phosphate to ADP-glucose conversion occurs in the cytosol and ADP-glucose is taken up by the amyloplast.

ADP-glucose pyrophosphorylase is a multimeric enzyme in plants, comprising two large and two small subunits. It is an interesting enzyme because mutations in the genes encoding either subunit have profound effects on starch synthesis: in maize, for example, the large subunit is encoded by the gene *Shrunken*-2 and the small subunit by *Brittle*-2, the gene names deriving from the appearance of the grain resulting from the loss of the gene's function.

ADP-glucose is the glucose donor for starch biosynthesis. Amylose is synthesised by starch synthase enzymes that are associated with starch granules, called granule-bound starch synthases, while amylopectin is synthesised by soluble starch synthases in conjunction with starch branching enzyme.

2.4.2 *Bioethanol production from starch*

The process of bioethanol production from starch is essentially the same as that from sucrose, except that additional preliminary steps have to be included in which the starch is broken down to simple sugars. This process also occurs in cereal grain and other starchy grains and tubers during germination and sprouting, and has been the subject of intense study; partly because of that, and partly because it is a key process in malting. In cereal grain, most starch is contained in the starchy endosperm, the largest organ in the seed and the raw material for the production of white flour. The cells in the endosperm are dead by the time the seed is mature, and the process of starch breakdown is initiated by the hormone gibberellic acid, which is released by the embryo and the aleurone (a thin layer of living cells surrounding the endosperm). When germination is initiated, the enzyme α-amylase is synthesised by the aleurone: this is an endoamylase that acts at random locations along the starch chain to yield shorter glucan chains (short chains of linked glucose units), facilitating the action of other degrading enzymes. Ultimately the products of this process are maltose (a disaccharide comprising two linked glucose units), maltotriose (a trisaccharide of three glucose units), and limit dextrin (a mixture of branched and unbranched glucans). Another enzyme, β-amylase, attacks the end of the glucan chain, cleaving off maltose molecules. Starch breakdown also involves a debranching enzyme (limit dextrinase), and the joint enzymatic capability of these enzymes is known as the grain diastatic power. This has been an important target trait for breeding barley and other cereal varieties for malting. In potato, starch degradation also involves starch phosphorylase (α-1,4 glucan phosphorylase), which acts on the terminal α-1,4-glycosidic bond to release glucose 1-phosphate.

The production of maltose from cereal grain starch reserves gives the process of malting its name. Malt, the product of malting, is an important feedstock in the brewing industry. During malting, grains of barley, rye or wheat are exposed to moist, warm conditions that induce partial germination and allow starch-degrading enzymes to be synthesised. Germination is halted before the starch is broken down completely by drying the partially-germinated grains in a kiln. The grains are then roasted at high temperatures, allowing the development of colour and flavour, crushed and milled into a grist, added to hot water, and subjected to a series of controlled, high-temperature 'stands' in a process known as mashing. It is during mashing that most starch breakdown occurs.

The malting process depends on the enzymes produced by the grain itself, and low amylase activity results in inefficient starch degradation and hazy beer with low alcohol content. Grain can also suffer from too much amylase activity before harvest. This is assessed by determining the Hagberg falling number, an international standard corresponding to the time it takes for a steel ball to fall through a slurry of flour and water that has been heated to release the starch and partially gelatinise it. If starch has been partly hydrolysed by α-amylase before harvest, the slurry is less viscous and the ball falls more quickly, giving a low Hagberg falling number. This is associated with poor grain quality and affects the price that the grain will command.

The traditional malting process is considered to be too complicated for the production of large quantities of bioethanol for transport fuel. Bioethanol production from cereal grains therefore generally uses a 'dry-grind' process (Figure 2.8) in which the entire kernel is ground with hammer mills into a coarse flour, then slurried with water to make a mash. Instead of relying on the intrinsic starch degradation enzymes in the grain, α-amylase is added to begin the starch breakdown process, and the mash is then cooked at more than 100°C. Mechanical shearing, together with the high temperature, breaks down large starch molecules. More

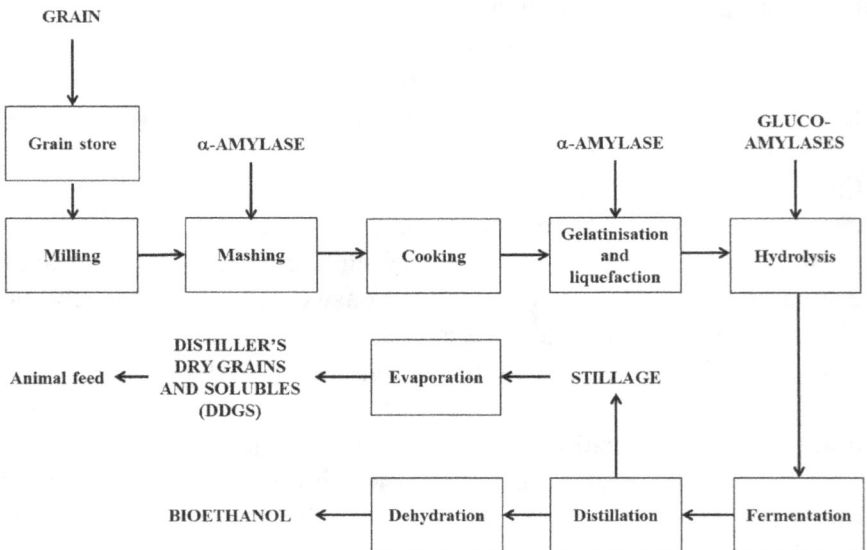

Figure 2.8   Flow-chart for the production of bioethanol and its animal feed co-products from grain starch by the 'dry-grind' process.

α-amylase is added to break the glucan chains down into maltotriose, maltose and limit dextrin in a process known as gelatinisation and liquefaction. The mash is then cooled and gluco-amylases are added to achieve saccharification, in which α-1,6-glycosidic bonds derived from amylopectin are hydrolysed and the short glucans are broken down further to produce glucose itself, ready for fermentation.

2.4.3 *Bioethanol from maize (corn) starch*

The biggest and most firmly established industry for the production of bioethanol from starch is in the US, and the feedstock of choice is maize (*Zea mays*) grain. Annual world maize grain production has now reached 844 million tonnes, overtaking rice (672 million tonnes) and wheat (651 million tonnes), making it the most produced crop commodity in the world in terms of tonnage. In addition, while most wheat and rice is used for food production, a lot of maize has traditionally been used for animal feed. Part of the business model for bioethanol production from grain is to use the starch to make bioethanol, and the high-protein residue to make a co-product to continue to supply the animal feed market. This stacks up, to some extent, because the animal feed market requires large amounts of quality protein. However, after a honeymoon period when this looked like a win-win situation, the industry has become controversial, mainly because of the amount of grain that it uses at a time when food security is causing increasing concern. The carbon intensity and sustainability of the industry and its true economics are also being called into question.

Currently, the maximum amount of bioethanol that can be made from maize grain is just over 400 litres per tonne, and the carbon intensity for the process when the bioethanol is made from US maize and used in the UK is $108\,g\,CO_2/MJ$, which is higher than that for petrol (Table 1.2). This means that greenhouse gas emissions are actually increased by 27% when substituting petrol with bioethanol from US maize starch in the UK. However, the carbon intensity for bioethanol from French maize starch is much lower, at 49 g CO_2/MJ; giving a greenhouse gas saving of 42% compared with petrol. The difference between the figures for the two countries could be explained in part by the additional energy requirement of transporting ethanol from the US to the UK, but the contrast again highlights how much the carbon intensity calculation can be affected by factors other than the basic process of converting the feedstock to the product.

As discussed in Section 1.5, the development of the US bioethanol industry has had considerable political support. Currently there is a subsidy of 51 cents per US gallon (13.5 cents per litre), the industry is protected from outside competition by the imposition of tariffs, and the Ethanol Mandate inscribed in the Energy Independence and Security Act called for a quadrupling in bioethanol production from its 2008 level of 9 billion gallons (34 billion litres) per year to 36 billion gallons (136 billion litres) per year by 2022. While new sources of ethanol such as cellulose (Section 2.5) are expected to contribute most of the increase, 15 billion US gallons (57 million litres) are expected to be produced from grain starch, most of it from maize. Bioethanol production was expanding rapidly anyway, even before the Ethanol Mandate, with annual growth from 1980 to the end of the last century of around 14%, accelerating to 25% in the first years of this century (Figure 2.9).

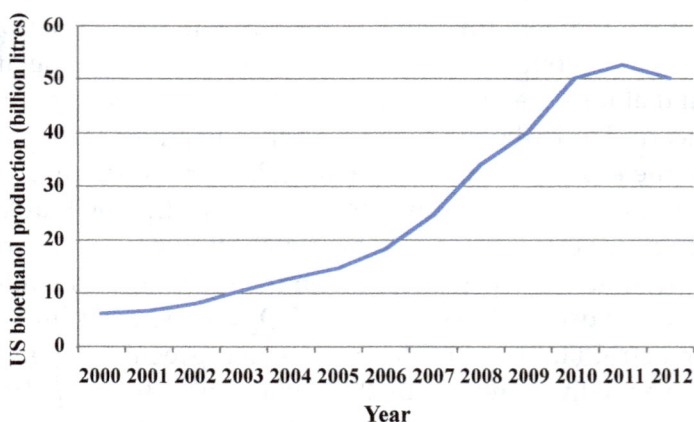

Figure 2.9   US bioethanol production, 2000 to 2012 (United States Energy Information Administration).

Production reached just under 53 billion litres in 2011, but fell back a little in 2012 to just over 50 billion litres. The ethanol is generally mixed with petrol in 5–10% blends to fuel ordinary petrol engines, but it is increasingly being used in 70 and 85% blends (E75 and E85) in flex-fuel vehicles.

The rapid increase in bioethanol production has inevitably resulted in an increase in the proportion of the maize harvest being used for fuel production. American maize provides one of the world's great harvests and, despite large domestic consumption, the US has traditionally been a major maize exporter. Concerns have therefore been raised over the effect of US bioethanol production on global food security and in 2012 the UN Food and Agriculture Organisation unsuccessfully called on the US government to suspend the Ethanol Mandate. Total US maize production and consumption from 2000 to 2012 are plotted in Figure 2.10, as well as the amount being used for bioethanol production and the amount being exported. Production rose steadily from 252 to 333 million tonnes between 2000 and 2009, which meant that the

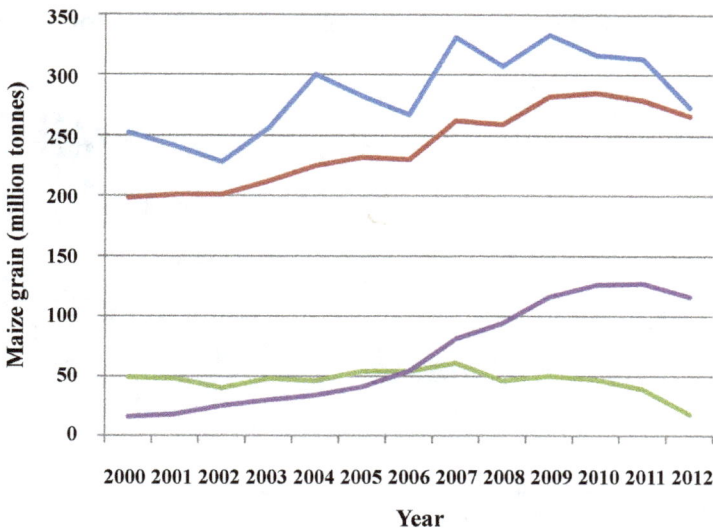

Figure 2.10   US maize grain harvest (blue), consumption (red), exports (green) and used for bioethanol production (purple), 2000 to 2012 (data from United States Department for Agriculture).

gap between production and consumption and consequently the amount being exported did not change greatly. This was despite the amount of grain being used for ethanol increasing from 18 million tonnes in 2000 to 127 million tonnes in 2011, an astonishing 42% of the total harvest.

It should be remembered that a lot of maize has always been used for animal feed, and that animal feed is a valuable co-product of the bioethanol industry, so the more than 100 million tonne increase in grain used for bioethanol has not increased total consumption by as much. Nevertheless, the price of grain has risen substantially this century (Figure 2.11), peaking at US\$333 per tonne in July 2012; more than double the price of US\$152 per tonne during the same month two years previously. The spike in price in 2012 was caused by a poor harvest, and, in that year, the gap between US production and consumption did close (Figure 2.10). The profitability of bioethanol production had already fallen from its peak of 80 cents per litre in 2006; the increase in grain price in 2012 led to that trend continuing, and was the cause of the slight drop in production that year. The price of grain had fallen back to just below US\$200 per tonne by late

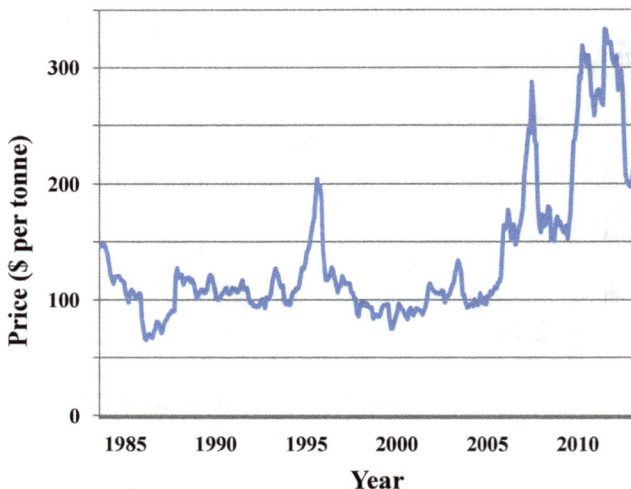

Figure 2.11   Maize grain price from 1984 to 2014 (Data from World Bank).

2013, on the back of a record harvest of 354 million tonnes: an increase of 40% over the year 2000 figure.

Clearly, some people will continue to question whether it is worth using 42% of the US maize harvest for fuel rather than food production. However, the fact that the industry is now well established, and enjoys strong political support means that the production of bioethanol from maize starch will continue until other feedstocks that can compete economically are available. It is notable that in 2013 the amount of grain being used for bioethanol may have been huge but it was almost entirely accounted for by the increase in grain production that had been achieved since the turn of the century. Nevertheless, continuing to meet the targets set by the US and other governments for bioethanol production will be a challenge, at a time when global demand for food is putting the supply chain under pressure. In developing countries, starch may provide as much as 80% of daily calorific intake, an important consideration if a significant proportion of it is to be used for non-food purposes. Starch is also an important renewable raw material for other industrial applications, including papermaking (starch typically makes up 8% of a sheet of paper), adhesive production, gypsum wallboard manufacture and textile yarns. The issue is certainly a complex one.

2.4.4 *Biotechnology*

Maize production in the US benefits from an established market for biotech (genetically modified) varieties and there is huge investment in this market from plant biotech companies. With over 40% of the crop now being used for bioethanol production, traits that are relevant to that end-use are now targets for improvement. The yield of sugar from starch is the most important cost determinant in the bioethanol production process, and Syngenta have already produced a biotech maize variety in which this trait is claimed to be improved. The variety contains a gene, *Amy797E*, from a thermophilic bacterium, *Thermococcales* spp., which encodes a highly thermostable α-amylase. It was deregulated (given the go-ahead to proceed to commercial development) by the US authorities in early 2011. Increasing the intrinsic amylase activity of the maize grain could reduce the need for adding enzymes during processing.

Another target for improvement is the quality of the high-protein co-product used for animal feed. The residual material remaining after distillation is referred to as the stillage, and includes protein, fibre and oil. This can be used directly as an animal feed but usually it is dried down to make DDGS, which is an important, high-protein ingredient for animal feed manufacture, and has long been a valuable co-product of the brewing and distilling industry. Its production consumes a lot of energy, but it is an essential part of the business plan for bioethanol production. Its value is, of course, dependent on its nutritional characteristics, and cereal grain and products, like DDGS, that are made from it usually have to be supplemented in animal feed because they contain insufficient amounts of the essential amino acid lysine. Renessen, a joint venture between Cargill and Monsanto, has produced a biotech maize variety with high lysine levels through expression of a gene from a bacterium, *Corynebacterium glutamicum*, encoding a lysine-insensitive dihydrodipicolinate synthase. Feedback inhibition of this enzyme is the major regulatory control for flux through the lysine biosynthesis pathway in plants, but the bacterial enzyme is not affected by lysine, allowing the amino acid to accumulate to high levels. The variety is currently being grown entirely for the bioethanol/animal feed market.

2.4.5 *Bioethanol from wheat: the United Kingdom experience*

Maize is not, of course, the only grain that could be used to produce bioethanol. Indeed, any starch-rich grain, tuber or root could provide the raw material for fermentation, and would be processed essentially in the same way as described for maize grain in the previous section. In Europe, the major grain crop is wheat (*Triticum aestivum*), with the European Union (with 28 Member

States) producing 132 million tonnes in 2012 at an average yield of 5.19 tonnes per hectare. The UK's contribution to that was 13.3 million tonnes, produced at 6.7 tonnes per hectare. This was a relatively poor yield for the UK, with production this century averaging 14.8 million tonnes at 8 tonnes per hectare. The poor yield was blamed on a wet spring and summer, and was one of the factors that brought about a cessation of production from April 2013 at one of the UK's bioethanol plants, run by Ensus at Wilton in the Teeside region of north-east England.

Ensus is owned by two US private equity funds, the Carlyle Group and Riverstone. It represents an investment of UK£300 million and was the first large bioethanol plant to be opened in the UK when it went into production in February 2010. The hiatus in production in 2013 was the second, following a previous shutdown from May 2011 until August 2012. In both cases the company cited 'adverse market conditions', caused in the more recent case by the poor wheat harvest of 2012, which affected the quality as well as the price of the wheat feedstock, on top of rising energy costs and a failure of the bioethanol price to rise in line with input costs. The workforce was not laid off, suggesting that the closure was not expected to be permanent, and production restarted in September 2013.

Wheat grain is a globally traded commodity. The size of the UK wheat harvest has little impact on the price of wheat, because UK production is such a small proportion of the total, which is over 650 million tonnes annually. The trend price for wheat on the London International Financial Futures and Options Exchange (LIFFE) towards the end of the 20th century was under UK£100 per tonne. However, severe droughts in two major wheat exporting countries, Australia in 2006–2007 and Russia in 2010, coupled with increasing demand, particularly from China, saw the price rise to a peak of just over UK£200 per tonne in December 2011. The price has fallen back, but very slowly, and for May 2014 was still UK£171 per tonne; mirroring the trend in world price shown in Figure 1.3. The long-term trend is expected to be upwards for the reasons described in Section 1.6. That is an issue for the future; Ensus' problems may be more to do with the quality of the feedstock and an inability to achieve the efficiency levels in their process that they had expected. Indeed, it has been reported that the company used imported maize grain to make up 20% of the feedstock in 2012, blaming the poor quality of wheat that was available. It is not clear what 'poor quality' means in this respect, but clearly the fermentability of the starch and the quality of the protein going into the animal feed co-product are possibilities.

The Ensus plant is designed to make over 400 million litres of bioethanol and 450 million tonnes of animal feed per year from approximately 1.2 million tonnes of wheat grain, making it one of the largest biorefineries in Europe. Animal feed is not the only co-product; the 300,000 tonnes of CO_2 produced by the plant is collected for use in the manufacture of carbonated drinks. A plant of similar size and cost (UK£350 million) has been built at Saltend in Hull, Humberside, also on the east coast of England, by Vivergo; owned jointly by Associated British Foods, DuPont and BP (British Petroleum). There seems to be more optimism about this venture, which went into production in 2013. It was officially opened on 8th July 2013 by the Right Honourable Vince Cable MP, a senior UK government minister, which suggests that the UK government has confidence in the venture, and in the biofuels business as a whole. The Vivergo plant will use 1.1 million tonnes of wheat per annum and produce 420 million litres of ethanol and 500,000 tonnes of animal feed, suggesting that the Vivergo operation is slightly more efficient than that of Ensus, with more ethanol and feed from less wheat. On the other hand, supplying CO_2 to the drinks industry does not appear to be part of the business plan, taking away one potential income stream that Ensus is exploiting.

Plans for a third major bioethanol facility have been proposed for Grimsby, also on the east coast of England, just south of the Vivergo plant on the opposite side of the Humber estuary. The plans were put forward by Vireol and the operation was to be financed in part by another venture capital group, Future Capital Partners. The plant would have cost approximately UK£200 million and been about half the size of the other two, producing annually

200 million litres of bioethanol, 175,000 tonnes of animal feed and 127,000 tonnes of CO_2 for the drinks industry from 500,000 tonnes of wheat. It was expected to start production in mid-2014, but the plans have been mothballed due to what the company calls 'delays in bringing forward the right regulatory framework in Europe', adding to the impression of a stuttering start for the bioethanol industry in the UK. Vireol is now focussing on operations in the US, although it claims to be optimistic about returning to the UK in the future.

The Ensus and Vivergo plants are in areas with an industrial tradition, and both are close to major ports through which ethanol can be exported into the European market and feedstock imported if necessary. Their location on the east coast of England also puts them close to the major wheat-producing areas of the UK. It will be interesting to see how the industry develops over the next few years, and what effect it has on wheat farming and the animal feed industry. The facilities will use feed wheat and produce a high-protein animal feed co-product, and the UK is currently short of protein while having an excess of starch so bioethanol production may actually balance that out. However, much of the protein that is currently imported for animal feed is in the form of soybean, and the low lysine content of wheat grain means that wheat protein cannot substitute for that entirely. This was discussed in Section 2.4.4 in the context of US maize, but there will be no biotech solution to that in Europe in the foreseeable future because of Europe's hostility to agricultural biotechnology.

Of the 15 million tonnes or so of wheat grain that have typically been produced in recent years in the UK, approximately 3 million tonnes has been exported, most of it feed wheat. It is this wheat that is likely to be used for bioethanol production instead. Nevertheless, if and when they are fully operational, the Ensus and Vivergo facilities will use 2.3 million tonnes of UK wheat, representing over 15% of the total UK wheat harvest in an average year. No doubt UK wheat farmers will be very happy to have another major use for their product, and given that the wheat price is dictated by what happens in other, larger wheat-producing countries the price of wheat may not be affected by UK operations alone. Nevertheless, it is no exaggeration to say that this will be the biggest revolution in the wheat market since the UK joined the European Union (or European Community as it was then) in 1973.

As with the maize bioethanol industry in the US, the development of the UK wheat bioethanol industry is partly politically driven. The European Union Directive 2003/30/EC established the goal of a 5.75% share of renewable energy in the transport sector by 2010, and Directive 2009/28/EC set a further goal of a minimum of 10% for every Member State to achieve by 2020 (Section 1.5). Europe currently consumes 134 billion litres of petrol per year, so 10% would represent 13.4 billion litres. The current carbon intensity figures for bioethanol from European wheat range from 59 g CO_2/MJ for French wheat to 65 g CO_2/MJ for UK wheat and 65 g CO_2/MJ for German wheat (Table 1.2), giving a greenhouse gas saving of 24–31% when used to substitute for petrol. However, recently EU officials have expressed concerns about the impact of biofuels on rising food prices, and the UK House of Commons Environmental Audit Committee has raised similar concerns, calling for a moratorium on biofuel targets. This has caused a degree of uncertainty in the industry. However, in 2013 the EU moved to help the industry by applying import duties to US bioethanol that, in the view of European officials, was being 'dumped' on the European market. This was one of the factors that led to the reopening of the Ensus plant in September 2013.

2.4.6 *Bioethanol from other grains*

Other starchy grain crops that could potentially be used for bioethanol production include barley (*Hordeum vulgare*), rye (*Secale cereale*), triticale (× *Triticosecale*), rice (*Oryza sativa*), oats (*Avena sativa*) and grain sorghum (*Sorghum bicolor*) (see Section 2.3.4 for sweet sorghum). Indeed, rye is a popular feedstock for bioethanol production in Poland, the Baltic States and Germany, and barley is used in Germany and Spain. A recent report from a UK agricultural consultancy, ADAS, highlighted the potential of triticale (a wheat–rye hybrid) as a bioethanol crop, claiming that switching to triticale from wheat would give an additional greenhouse gas saving of 14%, mainly due to triticale's lower nitrogen requirement.

Wheat has a dominant position in UK and European agriculture and it will be interesting to see if that changes as a result of the growth of the bioethanol industry. However, the bioethanol plants that have been built to date have been designed to use wheat as a feedstock, and penetration of the market by other cereals may depend on how easy it is to switch to a different grain. In the US, for example, grains other than maize have so far been considered uneconomical to use in bioethanol plants designed to use maize as a feedstock due to lower ethanol yields and poorer nutritional value of the animal feed co-product. This may not be the case, of course, when switching from wheat to rye, barley or triticale, but so far producers have used almost entirely wheat grain.

2.4.7 *Non-cereal sources of starch*

Potato (*Solanum tuberosum*) is the most important non-cereal source of starch and potato starch is already used in a number of industrial applications. However, if there is currently any use of potato starch for bioethanol production it is very small compared with that of cereal grains. This is because of the relatively high cost of the raw material, the yield of ethanol that is achievable and the quality of the animal feed co-product.

There is more interest in the use of tropical root crops such as cassava (*Manihot esculenta*) and sweet potato (*Ipomoea batatas*) because this would enable bioethanol production to become established in developing countries that currently are heavily dependent on imported transport fuels. Maize and sugar cane are produced in many tropical countries, of course, but cassava in particular will grow on soils that will not support economic cultivation of these crops and is productive in very low input farming systems with scarce water availability. Cassava also has the advantage of having extensive above-ground biomass that could also be used for electricity generation (Chapter 4). Its disadvantages are the relatively low starch content of the storage root (typically 20%) and the inefficiency of the farming, harvesting, transport and processing systems in many of the countries where it is grown. The food versus fuel question is also more critical in regions where food is already scarce, although, as for everywhere else, the issue is not a simple one and some would argue that revenue from bioethanol production could lift people out of poverty.

2.5 Second Generation Bioethanol from Cellulose and other Cell Wall Polysaccharides

2.5.1 *Plant cell walls*

While starch is the major storage carbohydrate of grains and tubers, other complex carbohydrates are integral to plant cell walls and therefore make up much of the biomass of stalks and leaves. As we have described, the first step in the predominant route for starch biosynthesis is the conversion of UDP-glucose to glucose 1-phosphate (Figure 2.7), and UDP-glucose is also used in the formation of cell wall polysaccharides. These include cellulose (Figure 2.6), a glucan very similar to amylose but comprising linear chains of glucose molecules joined by β-1,4 glycosyl linkages; (1,3;1,4)-β-glucan, which is similar to cellulose but has one β (1,3) linkage for every three or four β (1,4) linkages, xyloglucans, and xylans. Xyloglucan has a backbone of β (1,4)-linked glucose residues, some of which are substituted with (1,6)-linked xylose side-chains. It is a major cell wall polymer of dicotyledonous plants. Xylans are polymers of the pentose (5-carbon sugar) xylose, for which UDP-xylose is the xylose donor; UDP-xylose is produced from UDP-glucose by dehydrogenation to UDP-glucuronate, followed by decarboxylation. The major cell wall polymers of grasses are arabinoxylans, which are xylans in which some of the xylosyl units are subsituted with another pentose, arabinose, a stereoisomer of xylose. A proportion of the arabinose units are esterified with ferulic acid to form arabinofuranosyl side chains. Polymers comprising branched chains with more than one sugar type are sometimes collectively referred to as hemicellulose.

The other major carbohydrate component of plant cell walls is pectin, which is comprised of another complex group of polysaccharides. These include homogalacturonans, which are linear chains of α-(1-4)-linked galacturonic acid (an oxidized form of the sugar galactose); substituted galacturonans, which contain side-chains of xylose or apiose (another pentose) on a galacturonic acid backbone; and rhamnogalacturonans, which contain a backbone with repeating units of galacturonic acid and rhamnose (rhamnose is a deoxyhexose), with various sugar side-chains.

These complex polymers form a matrix with lignin, which is itself a complex polymer formed of aromatic alcohols. This matrix, called lignocellulose, is unusual because its composition and structure are variable. Lignocellulose is the most abundant renewable raw material on Earth. Approximately 10% of the carbohydrate in cereal grain is in the form of lignocellulose, so its exploitation could lead to significant increases in ethanol yield from grain. Perhaps more importantly, it makes up most of the stems and leaves, to the extent that the terms lignocellulose and biomass are sometimes used interchangeably. Exploitation of this material for energy would make cereal straw a commodity with value and, of course, would enable energy to be generated from a part of the plant that is not eaten. There is already interest in using straw as biomass for electricity generation or as a feedstock for biogas production (Chapters 4 and 5). However, it may also be possible to make bioethanol from it.

In theory, the cellulose and hemicellulose components of lignocellulose could be broken down into fermentable sugars by cellulase and hemicellulase enzymes. In practice, they are locked away in an impenetrable matrix with lignin, and their extraction from the matrix requires one or more of a variety of pre-treatments. These include hydrolysis with hot water or steam, which releases the hemicellulose. Steam is also applied under pressure, followed by transfer of the material to a low-pressure vessel, causing rapid expansion of the material and making it more penetrable. Another method combines water and air to oxidise the lignin to low molecular weight carboxylic acids and alcohols. Several chemical pre-treatments are used, including dilute sulphuric acid, which extracts the hemicellulose and makes the remaining polymer more accessible to cellulase, as well as ammonia, lime and ionic liquids. Lastly there is biological pre-treatment, using fungi and/or bacteria to degrade the lignin. All of these have a cost associated with them and none are 100% effective. Nevertheless, the technology continues to develop and cellulosic bioethanol is expected to make a major contribution to filling the gap between how much bioethanol can be produced from current feedstocks and the targets set by governments for the future.

2.5.2 *Bioethanol from algal cell wall polysaccharides*

Algae are aquatic, photosynthetic organisms. They include green algae, which can be regarded as single-celled plants, and blue-green algae (cyanobacteria), which are photosynthetic bacteria. These single-celled algae are also known as microalgae or phytoplankton. Algae also include seaweeds, or macroalgae, which are multicellular plants but lack the true stems, leaves and roots of higher plants.

Algae have been used as a source of food, food supplements, chemicals, pharmaceuticals, cosmetic ingredients and animal feed for centuries. Interest in their potential for biofuel production arises mainly because of their high yields, which may be many times those of even the most productive land plants. Many are edible, meaning that it may also be feasible to produce an animal feed co-product.

Seaweeds are already harvested from the wild or cultivated in near-shore plantations (marine culture or mariculture). Cultivated species include *Ascophyllum* (Norwegian kelp), *Eucheuma and Kappaphycus* (edible red seaweeds used as a source of carrageenan, which are sulphated polysaccharides used as gelling and stabilising agents in food processing), *Fucus* and *Laminaria* (brown, shoreline seaweeds), *Gracilaria* (a red seaweed that produces agar, a gel used widely for plant, animal and microbial cell culturing), and *Porphyra* (a red, shoreline seaweed used to make laverbread and other foods). The annual yield of seaweed is estimated to be just under 20 million tonnes wet weight, with most coming from Asian mariculture.

Some microalgae are also already cultivated, including the green algae, *Chlorella*, *Dunaliella* and *Haematococcus*, and *Spirulina*, which actually comprises two species of blue-green algae, *Arthrospira platensis* and *Arthrospira maxima*, that are found in tropical and sub-tropical lakes. *Spirulina* is used as a food supplement in many countries. Other traditional uses include the production of ingredients for cosmetics and dyes.

Traditional algal cultivation is in open ponds or raceways, in which the algae use sunlight and atmospheric carbon dioxide to photosynthesise (an autotrophic system). These systems are relatively cheap and require little input, but are prone to contamination by the wrong algae, fungi and bacteria. It may also be impossible to attain growth rates and cell densities in these systems sufficient for commercial biofuel production. There are many projects around the world to develop efficient closed systems in which light, carbon dioxide and nutrients are controlled to give optimal growth rates and cell densities, and undesirable organisms are excluded. Such heterotrophic systems would have to be supplied with carbon dioxide (bioreactors) and possibly artificial light (photobioreactors), adding to the cost and carbon intensity.

A leader in the development of bioethanol production from algae is the company Algenol, based in Fort Myers, Florida. Algenol describes itself as a global, industrial biotechnology company that is commercialising its patented algae technology platform for production of bioethanol and other biofuels. Algenol's system uses blue-green algae grown in bioreactors covering 16 hectares. Ethanol is produced directly from the algae and the remaining biomass is then used to make other liquid fuels. The company claims an 85% conversion rate of CO_2 feedstock into product, a carbon saving of 60% and net production of approximately 75000 litres per hectare (about 20 times the yield of bioethanol from maize).

Ethanol is the major product, making up approximately 85% of the total, with biodiesel 6.2%, petrol 4.7% and jet fuel (typically comprising hydrocarbons containing from 6 to 16 carbon atoms) 4.0%. The company is aiming to go into commercial production in 2015.

A feasibility study on the exploitation of seaweed for bioethanol has been undertaken by the Danish Technological Institute, in collaboration with Aarhus University, Technical University of Denmark, National University of Ireland in Galway, the University of Hamburg and University of Sienna, Italy. The study looked at two *Laminaria* algae, *Saccharina latissima* and *Laminaria digitata*, the first grown on 10 km of seeded lines, the second harvested from natural populations. The cultivated *Saccharina latissima* was grown for 7 to 8 months and yielded two tonnes wet weight per 2 km line. This sort of cultivation involves the collection of spores, which are seeded onto strings where they germinate into small plants. The plants are then placed on long lines and transferred to the sea.

The major polysaccharides in *Laminaria* algae are those present in the cell walls, mainly cellulose and alginate. Cellulose (Figure 2.6) has already been described, while alginate is a linear, anionic polysaccharide comprised of D-mannuronate ($C_6H_9O_7$) (M) and L-guluronate ($C_6H_{10}O_7$) (G) residues, in blocks either of residues of the same type, that is MMMMMMMM or GGGGGGGG, or alternating types, MGMGMGMG. Cellulose can be hydrolysed by treatment with cellulase (Section 2.5.1) and alginate by treatment with another enzyme, alginate lyase. This generates simple sugars for fermentation. The researchers obtained approximately 7 litres of ethanol from 100 kg wet weight of seaweed.

Clearly, it would take a huge operation to make sufficient bioethanol from either cultivated or wild seaweed to contribute significantly to liquid transport fuel production, and the harvesting of wild populations would have to be carefully risk-assessed for its environmental impact. However, the researchers claimed that their study showed that the process was feasible. As is the case with the novel crops that will be discussed in subsequent chapters, cultivating algae for the production of high-value niche products is one thing, but producing algae on the scale required for biofuel production is another; and the lack of experience of large-scale production is a major constraint on the development of the industry.

2.6 Butanol

Butanol (butyl alcohol) is a four-carbon alcohol, formula C_4H_9OH. It has four possible isomers: 1-butanol, 2-butanol, *iso*-butanol (2-methyl-1-propanol) and *tert*-butanol (Figure 2.12), but some bio-fuel specifications exclude *tert*-butanol. The longer carbon chain of butanol makes it less polar than ethanol, meaning that it is easier to mix with petrol. It also has an energy rating of 29.2 MJ per litre, much closer to that of petrol (32 MJ per litre) than ethanol (19.6 MJ per litre). This means that blends of butanol and petrol with up to 85% butanol can be used in unmodified petrol engines.

Like ethanol, butanol can be made by microbial fermentation of sugars, but the established microbe for the purpose is not a yeast but a bacterium, *Clostridium* spp. The process produces acetone and ethanol as well as butanol and hence is called the ABE (acetone, butanol and ethanol) process. It is inefficient and had almost disappeared in the face of competition from synthetic production

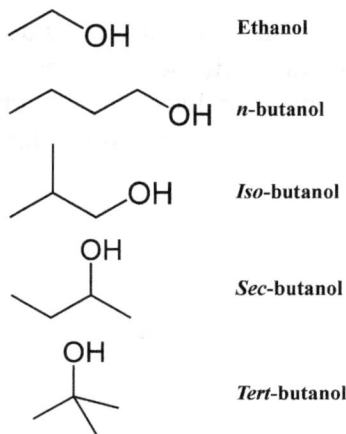

Figure 2.12 Line diagrams showing the structures of different forms of butanol (C_4H_9OH). Ethanol is also shown for comparison.

processes until the recent interest in the production and use of butanol as a biofuel.

Possible solutions are biological and chemical. Strains of *Clostridium* have been developed that convert cellulosic biomass as well as sugars to butanol, greatly increasing yield. The bacterium has also been modified to reduce the production of acetone and ethanol in favour of butanol. Genetically modified yeasts that make butanol instead of ethanol have also been developed. Chemical solutions to the problem include the development of catalysts that convert ethanol to butanol, allowing butanol production to be added to existing bioethanol plants.

Commercially, the leading players appear to be Butamax, a joint venture between BP and DuPont, and Gevo. Gevo has developed what it calls Gevo Integrated Fermentation Technology, incorporating advanced catalytic and separation technologies. It commenced production at a converted bioethanol plant in Luverne, Minnesota, US in 2012. Butamax technology involves 'recombinant microbes' and advanced separation technologies. In 2013, the company began to work with Highwater Ethanol LLC to convert Highwater's ethanol plant in Lamberton, Minnesota, to make butanol.

Biobutanol production is still a very small industry compared with bioethanol. However, given butanol's advantages over ethanol as a fuel to blend with petrol, biobutanol could replace bioethanol in the transport fuel sector if the challenge of developing production processes that match those for ethanol in terms of cost and efficiency can be met.

3 BIODIESEL

3.1. Introduction

The second transport fuel derived from crops is biodiesel, which is produced from plant oils by transesterification with methanol, or, more rarely, ethanol, to make fatty acid methyl (or ethyl) esters (FAMEs). The principle components of plant oils and animal fats are triacylglycerols, which are made up of three fatty acid molecules attached to glycerol (Figure 3.1). Fatty acids comprise a carboxylic acid (COOH) group at the end of a hydrocarbon chain, which is generally regarded as including at least eight carbon atoms (the 18-carbon stearic acid is shown in Figure 3.1). Naturally occurring fatty acids are synthesised from the two-carbon acetyl group of acetyl-CoA (Figure 3.1), and therefore have an even number of carbon atoms.

Different fatty acids are distinguished not only by chain length but also by the number and position of double bonds between the carbon atoms in the chain. Carbon atoms joined by double bonds are described as unsaturated, while those that are joined by single bonds are described as saturated. Fatty acids in which all of the carbons are joined by single bonds are similarly called saturated fats, while those with a single double bond are called monounsaturates, and those with more than one are called polyunsaturates. Plant oils contain a variety of fatty acids with different chain

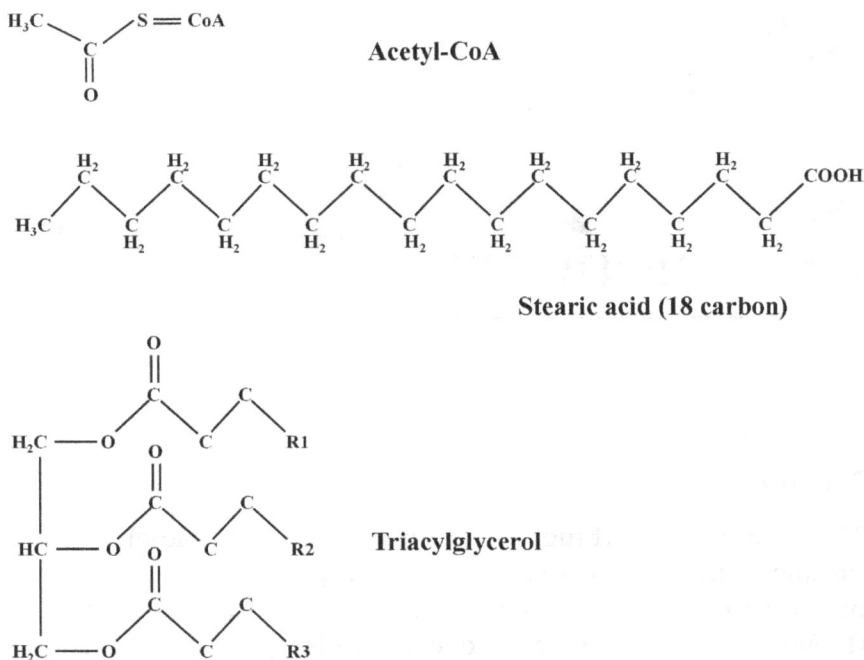

Acetyl-CoA

Stearic acid (18 carbon)

Triacylglycerol

Figure 3.1 Diagrams showing the structure of acetyl-CoA, the 18-carbon saturated fatty acid, stearic acid, and triacylglycerol (R1, R2 and R3 represent alkyl chains derived from fatty acids).

lengths and degrees of saturation (Table 3.1) and the various properties of oils from different plants are determined by their differing fatty acid contents. Animal fats, in contrast, contain mostly saturated and monounsaturated fatty acids.

The number of carbon atoms and double bonds in the fatty acid chain is given as a ratio; for example, oleic acid is an 18-carbon fatty acid with a single double bond (a monounsaturate), so the ratio is 18:1. The position of the first double bond in the chain is given in the form n-x, where x is the position of the first unsaturated carbon with respect to the methyl (omega) end of the molecule; so, for oleic acid, this number is n-9. Oleic acid (18:1, n-9) is also known as an omega-9 fatty acid for this reason. Oleic acid is the

Table 3.1 Some plant fatty acids.

Name	Length (carbon atoms)	Saturated, monounsaturated or desaturated	Position of double bonds	Comment
Caprylic acid	8	Saturated	Not applicable	
Capric acid	10	Saturated	Not applicable	
Lauric acid	12	Saturated	Not applicable	
Myristic acid	14	Saturated	Not applicable	
Palmitic acid	16	Saturated	Not applicable	
Stearic acid	18	Saturated	Not applicable	
Oleic acid	18	Monounsaturated	n-9	Omega-9
Linoleic acid	18	Polyunsaturated	n-6, n-9	Omega-6
α-Linolenic acid	18	Polyunsaturated	n-3, n-6, n-9	Omega-3
γ-Linolenic acid	18	Polyunsaturated	n-6, n-9, n-12	Omega-6
Erucic acid	22	Monounsaturated	n-9	Omega-9 Toxic

major constituent of olive and oilseed rape oil. Other well-known plant fatty acids include lauric acid (12:0) and palmitic acid (16:0), which are the prevalent fatty acids in coconut and palm oil, stearic acid (18:0), a major component of cocoa butter, and linoleic acid (LA) (18:2, n-6), which is found in sunflower and maize oil and makes up about 20% of oilseed rape oil. Tallow (rendered animal fat) is made up predominantly of stearic acid and oleic acid.

The fact that fatty acids have a non-polar hydrocarbon end terminating in a methyl (CH_3) group and a polar carboxylic acid (COOH) group at the other end means that fatty acids can interact with both water and fatty or greasy substances. Hence they had a variety of industrial as well as food uses before the expansion of the biodiesel market, including the manufacture of products such as soaps, detergents and shampoos, as well as lubricants. However, these non-food uses are now dwarfed by biodiesel production, which has expanded rapidly this century to reach 21.75 billion litres per year by 2011 (Figure 3.2).

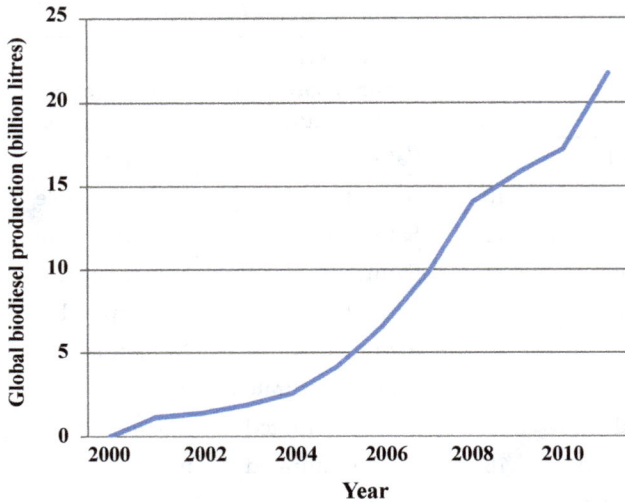

Figure 3.2   Global annual biodiesel production, 2000–2011 (Data from UN Food and Agriculture Organisation).

As with bioethanol, the US is the biggest producer (Table 3.2), but it does not dominate the biodiesel market in the way it does the bioethanol market; its annual production of 3.66 billion litres in 2011 made up only 17% of the global total. In contrast to bioethanol production, the industry is also well established in Europe: Germany produced 3.02 billion litres in 2011, giving it 14% of total production and making it the second largest producer, while France produced 1.97 billion litres, putting it fifth in the global table. Third and fourth places were taken by Brazil and Argentina.

3.2. Oil Synthesis in Oilseed Crops

In seeds, the carbon that is required for fatty acid synthesis comes from sucrose imported from photosynthesising leaves. The first step is the hydrolysis of the sucrose by invertase, to form glucose and fructose (Figure 2.3), followed by the phosphorylation of these hexoses to produce hexose phosphates. The hexose phosphates

Table 3.2 World biodiesel production 2011.

Country	Production (billion litres *per annum*)
United States	3.66
Germany	3.02
Argentina	2.75
Brazil	2.67
France	1.97
Indonesia	1.16
Spain	0.70
Italy	0.65
Netherlands	0.56
Belgium	0.50
China	0.45
Poland	0.44
Colombia	0.42
Republic of Korea	0.37
Austria	0.36
Portugal	0.32
Malaysia	0.26
Czech Republic	0.24
Canada	0.16
Denmark	0.16
Hungary	0.16
Greece	0.14
Australia	0.09
Lithuania	0.09
Romania	0.09
Serbia	0.09
Other	0.53
Total	21.75

Source: UN Food and Agriculture Organisation.

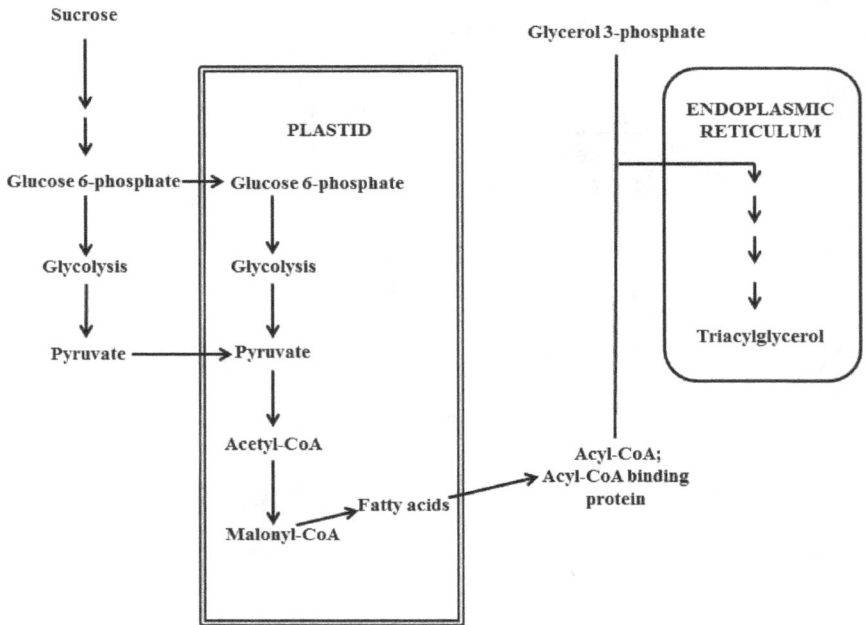

Figure 3.3   Diagram showing the compartmentation of fatty acid synthesis and assembly into triacylglycerols.

then enter glycolysis and are converted through multiple steps to the three-carbon molecule, pyruvate. This was described previously in the context of ethanol production from sucrose by yeast (Section 2.2). Here, though, the pyruvate is converted to acetyl-CoA (CH_3-CO-S-CoA) in plastids by a multi-protein complex called the pyruvate dehydrogenase complex. The plastids either import pyruvate or generate it through glycolysis themselves (Figure 3.3).

Acetyl-CoA has many potential uses in the cell, and the first committed step in fatty acid synthesis is its carboxylation to form malonyl-CoA (COO^--CH-CO-S-CoA) (Figure 3.4). This reaction is catalysed by the enzyme acetyl-CoA carboxylase and requires one molecule of bicarbonate (CHO_3^-) and one molecule of ATP. Malonyl-CoA is used in fatty acid biosynthesis by an enzyme

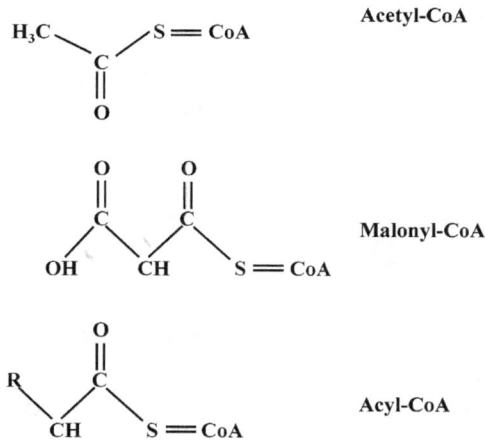

Figure 3.4 Diagrams showing the structures of key compounds in fatty acid biosynthesis, acetyl-CoA, malonyl-CoA and acyl-CoA.

complex called fatty acid synthase. The process is multi-step, involving transfer of the acyl group (RCO-, where R is a hydrocarbon chain), the sequential addition of two-carbon units and finally termination of the reaction. Another protein, acyl carrier protein (ACP), is involved in the intermediate reactions.

The reactions are catalysed by four enzymes: β-ketoacyl-ACP synthase (KAS), β-ketoacyl-ACP reductase, β-hydroxyacyl-ACP dehydrase and enoyl-ACP reductase. The predominant products are the 16- and 18-carbon saturated fatty acids, palmitic and stearic acid (Figure 3.1). These are exported from the plastid and interact with coenzyme A (CoA) and acyl-CoA binding protein (Figure 3.3). The resulting complex is transported to the endoplasmic reticulum, where the fatty acids are assembled via lysophosphatidic acid, phosphatidic acid and diacylglycerols into triacylglycerols. During or after this process the fatty acid chains may be lengthened and/ or desaturated by elongases and desaturases.

3.3. Biodiesel Manufacturing

Triacylglycerols (Figure 3.1) are esters of fatty acids and glycerol, esters being compounds formed by the condensation of a fatty acid with an alcohol and having the general formula:

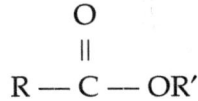

$$\begin{array}{c} O \\ \| \\ R - C - OR' \end{array}$$

where R and R' are different hydrocarbon groups. The most common method of biodiesel manufacture (Figure 3.5) is through transmethylation (also known as Fischer esterification or transesterification with methanol), which produces individual fatty acid methyl esters (FAMEs) with glycerol as a by-product. The reaction is catalysed by sodium hydroxide and has the overall equation:

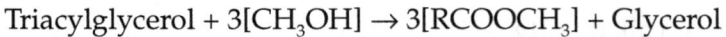

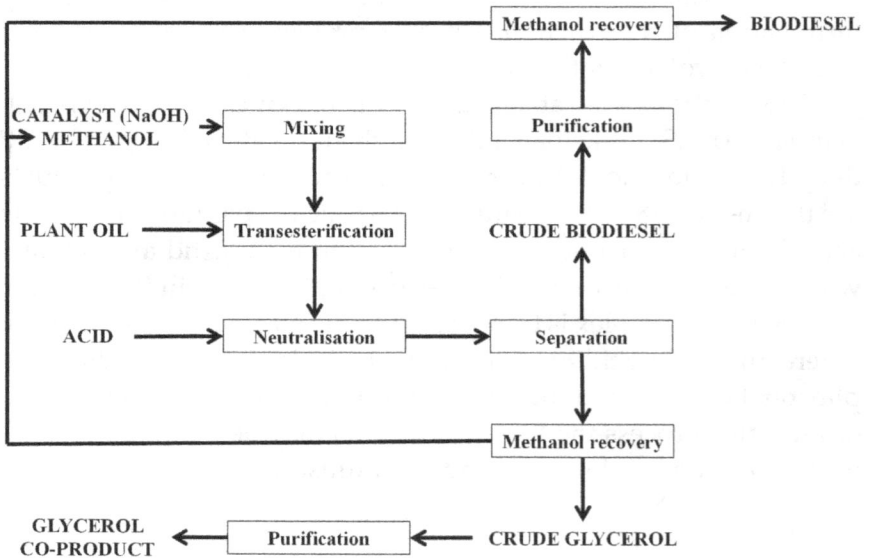

$$\text{Triacylglycerol} + 3[CH_3OH] \rightarrow 3[RCOOCH_3] + \text{Glycerol}$$

Figure 3.5   Flow diagram representing the biodiesel manufacturing process.

Stearic acid

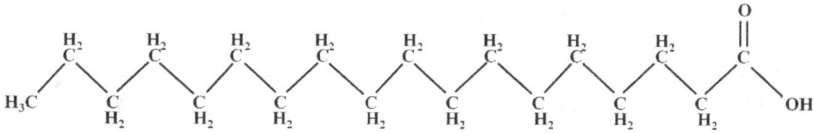

Stearic acid methyl ester

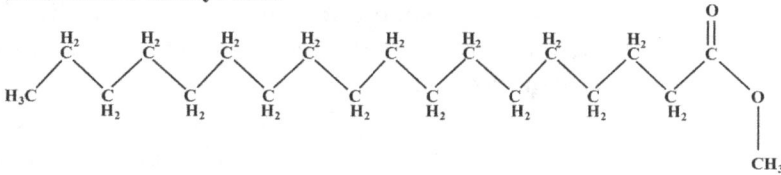

Stearic acid ethyl ester

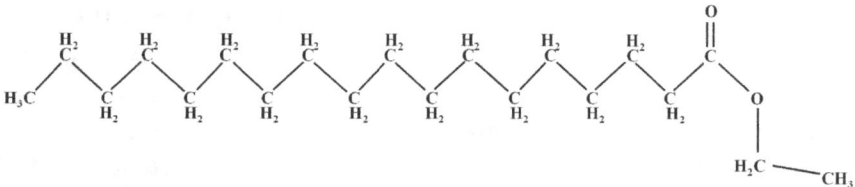

Sodium stearate (soap)

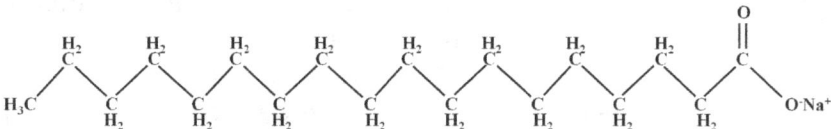

Figure 3.6 **Stearic acid, a common 18-carbon, saturated fatty acid, and the methyl- and ethyl-esters that are derived from it during the manufacture of biodiesel. Sodium stearate, which is the soap derived from stearic acid and a contaminant in biodiesel, is also shown.**

Ethanol may be used in the reaction instead of methanol to produce fatty acid ethyl esters ($RCOOCH_2CH_3$). Stearic acid, its methyl and ethyl esters, and, for comparison, sodium stearate (the soap derived from stearic acid) are shown in Figure 3.6.

The by-product, glycerol (a sugar alcohol) has a variety of uses in the food and cosmetics industries, but the amount being produced by the biodiesel industry means that there is a glut of it and other uses are being explored. One of these is to use it as a feedstock for biobutanol production (Section 2.6).

3.3.1. *Biodiesel quality compared with petroleum-based diesel*

The combustibility of diesel fuel during compression ignition is rated using an industry standard called the cetane number. Higher cetane diesel fuels have shorter ignition delay periods than lower cetane fuels, giving a smoother and quieter drive. The cetane number is derived by burning the diesel under standard test conditions and determining which mixture of two reference fuels, cetane (also known as hexadecane) and *iso*-cetane (also known as heptamethyl-nonane), gives the same ignition delay. Cetane (Figure 1.4), which is a 16-carbon alkane (saturated hydrocarbon), chemical formula $C_{16}H_{34}$, ignites very easily under compression and has been assigned a cetane number of 100. *Iso*-cetane, which has the same chemical formula as cetane, but is highly branched and ignites much less easily, has a cetane number of 15. The original second reference fuel was 1-methylnaphthalene, which was assigned a cetane number of zero, but *iso*-cetane is now used instead because it is cheaper and safer to handle.

Clearly, the methyl or ethyl ester composition of biodiesel reflects the fatty acids present in the oil from which it is produced, and therefore the properties of the biodiesel. In other words, not all biodiesels are the same, because they are derived from different feedstocks. Generally speaking, however, biodiesel has a higher cetane rating than diesel made from petroleum, with biodiesel made from plant oils typically having a cetane number between 46 and 52, while the cetane number of biodiesel from animal fat ranges from 56 to 60. Modern diesel engines operate well on fuel with a cetane number from 40 to 55. In the European Union, the current standard for diesel set out in Directive 2009/30/EC is a cetane number of 51. Premium diesel fuel can have a cetane number as high as 60, but this may be achieved by using additives that facilitate ignition. In the US the standard in most states is 40, but in California it is 53.

The cetane number is not the only consideration for diesel, the others being calorific value, lubricity (a measure of lubricating properties: important for fuel pumps and injectors), performance in the cold and sulphur content. The calorific value of biodiesel is typically between 37 and 38 MJ/kg, compared with 40 to 41 MJ/kg for petroleum-derived diesel; so vehicles driven on biodiesel will achieve slightly less distance per litre. On the other hand, biodiesel gives better lubricity and has a relatively high flashpoint (fuel ignition temperature) compared with petroleum-based diesel, reducing the risk of unintended ignition. The flashpoint for soybean-based biodiesel, for example, is over 100°C, while that of petroleum-based diesel is typically 71°C.

Biodiesel also scores well for low sulphur content, and this is now an important regulatory issue. Almost all diesel fuel sold in the European Union and North America is now classed as ultra-low-sulphur diesel, although the definition of ultra-low-sulphur diesel is changing as the regulations are tightened; a process that is ongoing as governments attempt to reduce the emissions from diesel engines. In the European Union, the 'Euro V' standard stipulates a sulphur content of 10 parts per million (ppm), and Germany has introduced tax incentives to encourage the development and use of essentially sulphur-free fuel. In the US, ultra-low-sulphur diesel has been the standard for road vehicles since 2006 and will be required for all other diesels from 2014, but is defined as having 15 ppm sulphur or less, not 10 ppm. The inclusion of biodiesel has facilitated the development of fuels that meet these standards, partly because of the low sulphur content of biodiesel and partly because of its greater lubricity, which makes up for the lower lubricity of ultra-low-sulphur diesel from petroleum. Biodiesel is also more biodegradable than petroleum-derived diesel, and this is particularly important for marine applications.

A parameter that biodiesel scores poorly on is performance at low temperatures. Oils/fats with a high proportion of saturated fatty acids, such as tallow (rendered animal fat) and palm oil, have a relatively high melting temperature and are semi-solidified at temperatures typical of temperate zones such as northern Europe and North America, even in the summer. The same is true of the biodiesels derived from them, unless they are blended with other diesels or additives. Used unblended, these biodiesels can clog fuel lines and filters in a vehicle's fuel system, or even become so solid that they cannot be pumped.

Cold temperature performance is defined using the cloud point, which is the temperature at which crystals start to form and the diesel starts to appear cloudy; and the cold filter plugging point, which is the temperature at which the crystals become too large to pass through a 45 µm filter. Most countries set standards for cloud point and cold filter plugging point that reflect the minimum temperatures that the diesel is likely to have to perform in within that country. Even within the European Union, there are differences in the requirements between countries, with the UK, for example, setting a cold filter plugging point standard of −5°C in summer and −15°C in winter, while the standard in Spain is 0°C in summer and −10°C in winter. This means that a motorist who fills up with diesel in Spain before the end of September and then takes a car ferry to the UK, may be driving in the autumn in the UK with diesel that will not perform at 0°C, a temperature that is quite likely to occur overnight. Most motorists are unaware that the EU does not have a single standard.

The poor cold performance of some biodiesels can be addressed by blending with petroleum-based diesels, or with biodiesels derived from alternative feedstocks. These are discussed in more detail in the following sections; but while tallow and palm oil are solid or nearly solid at the ambient temperatures that might be expected in temperate countries, the melting point of biodiesel from rapeseed oil is −10°C. In the US, diesels may be blended with kerosene (a petroleum fraction containing a variety of straight-chain and aromatic hydrocarbons with between 6 and 16 carbon atoms, called paraffin in the UK), which has a cold filter plugging point of −40°C. A range of additives that improve cold temperature performance are also used.

3.3.2 *Commercial blends*

Biodiesel can be used in diesel engines as a pure fuel, but in the US is usually mixed with petroleum-based diesel to give blends containing from 2% biodiesel (given the classification B2) to 20% (B20). Higher biodiesel blends are used in Europe, with 100% bio-diesel (B100) available in some countries. Solidification (gelling) at low temperatures is one reason why biodiesel may not be used on its own. The other reason is supply. The two countries that top the biodiesel production list, for example, are the US, with 3.66 billion litres per year, and Germany, with 3.02 billion litres per year (Table 3.2); but even in these countries this represents only 1.6% and 4.7%, respectively, of the total diesel fuel that is consumed, which is 232 billion litres per year in the US and 64 billion litres per year in Germany (data from US Energy Information Administration).

3.4 Biodiesel Feedstocks

3.4.1 *Soybean oil*

Soybean (*Glycine max*) is a legume that has been used as a food crop in China for 5000 years. It is an annual plant and like all legumes it fixes its own nitrogen from the atmosphere and so does not require nitrogen fertilizer. Soybean was first grown in Europe and the Americas in the late 18th century, and it was the late 19th century before it became a major agricultural crop, becoming established at that time in the US, Brazil and Argentina.

By the mid-20th century, soybean meal made from crushed beans was becoming a major component of animal feed, and the oil that was extracted as the beans were crushed was being used as a cooking and food oil. Up to that time, cereal grain had been the major raw material for the production of animal feed, but, as discussed in Section 2.4.4, cereal grain has relatively low concentrations of the amino acid lysine. Animals cannot synthesise lysine, and must obtain sufficient of it in their diet (hence lysine is one of those amino acids that is described as 'essential'). Soybean, like other legumes, contains significantly more lysine than cereal grain (soybean meal contains approximately 3% lysine, while maize grain, for example, contains less than 0.3%). The demand for animal feed grew fast as the populations of North America and Europe became more prosperous and meat consumption increased. By the 1970s, soybean cultivation in the US had risen to cover more than 20 million hectares, but US farmers were still unable to satisfy demand as the European animal feed industry began to use imported soybean in its formulations. The resulting price increase led to large increases in soybean production in Brazil and Argentina. Today, despite soybean's Asian origins, the Americas account for almost 90% of total production (Table 3.3).

Interest in soybean oil as a biodiesel feedstock did not begin until the 1990s. The National Biodiesel Board was founded in the US in 1992, with soybean-grower and federal funding. Further impetus was provided by tax incentives that were adopted in 2005 to favour biodiesel use, and this led to a huge increase in US biodiesel production, from only 200 million litres in 2005 to 3.66 billion litres in 2011. Most US biodiesel is produced from

Table 3.3   Soybean production in 2012 in countries producing more than 1 million tonnes *per annum*.

Country	Production (million tonnes *per annum*)
United States	89.5
Brazil	87.5
Argentina	54.0
China	12.2
India	11.0
Paraguay	8.1
Canada	5.2
Uruguay	3.5
Ukraine	2.8
Bolivia	2.3
Russian Federation	1.6
EU-27	1.2

Source: US Department of Agriculture.

soybean oil. There has been similar political intervention in Brazil, with the blending of biodiesel with all petroleum-derived diesel becoming mandatory in 2008, and the minimum biodiesel component now being set at 5%. Brazil and Argentina made 2.75 and 2.67 billion litres of biodiesel, respectively, in 2011; almost all from soybean oil.

Soybean oil is composed primarily of five fatty acids: the saturated fatty acids, palmitic acid (16:0) (typically 13%) and stearic acid (18:0) (4%); the monounsaturate, oleic acid (18:1) (18%); and the polyunsaturated fatty acids, linoleic acid (18:2) (55%) and α-linolenic acid (18:3) (10%). The high proportion of polyunsaturated fatty acids in the oil means that soybean biodiesel performs relatively well in cold temperatures, and soybean biodiesel has a cloud point of −10°C. The downside to the presence of a high proportion of polyunsaturated fatty acids is that the oil is prone to oxidation, which affects the oil's long-term storage and can cause engine gumming, although this can be addressed with antioxidants and synthetic additives. Soybean oil biodiesel also slightly increases emissions of nitrous oxides compared with petroleum diesel.

The major limitation on further growth of the US biodiesel industry is the availability of soybean oil. In the US, production has increased from around 5 million tonnes in 1983 to 9 million tonnes in 2013 (Figure 3.7). Soybean is grown on just under 30 million hectares of land in the US, but seed yield is only 3 tonnes per hectare, and oil only a tenth of that (less than 350 litres per hectare). Biodiesel production already takes more than a third of US soybean oil, yet, as discussed above, represents less than 2% of total US diesel usage. The increasing demand has led to a steep rise in price (Figure 3.8), from its long-term trend of around US$400 per tonne to over US$1400 per tonne in 2008; despite the increase in production. The price has fallen back from its 2008 peak, but continues to be over US$800 per tonne. The use of soybean oil for biodiesel is not the only driver for increasing demand, but it is clearly a factor.

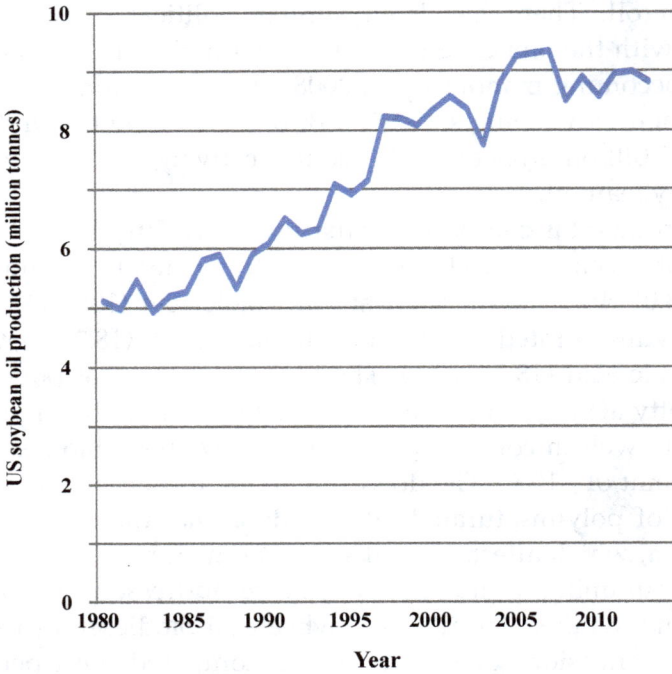

Figure 3.7   US soybean oil production from 1980 to 2013 (Data from US Department of Agriculture).

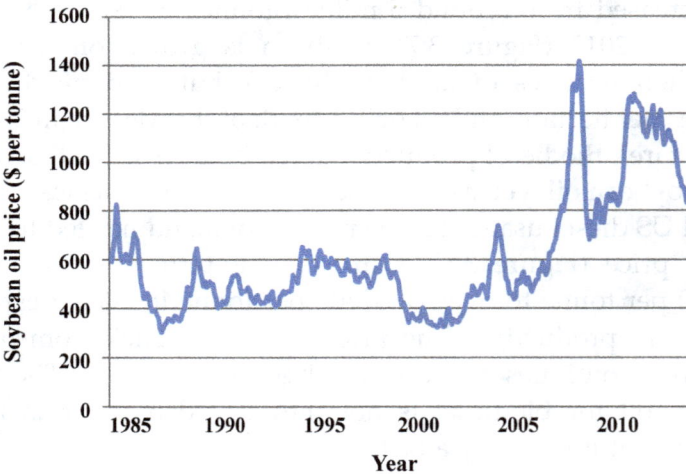

Figure 3.8   Soybean oil price from 1983 to 2013 (Data from World Bank).

3.4.2 *Oilseed rape (canola) oil*

In Europe and other temperate zones, the principal biodiesel feed-stock is oilseed rape oil, which makes up about 13% of total vegetable oil production globally. Oilseed rape production actually involves four closely related species: *Brassica napus*, *Brassica rapa*, *Brassica juncea* (sometimes called Indian mustard) and *Brassica carinata* (sometimes called Ethiopian mustard). The common names of swede rape and turnip rape are sometimes applied to *Brassica napus* and *Brassica rapa*, but these seem to be falling out of use. *Brassica napus* dominates world production, particularly in Europe and China, but *Brassica rapa* is widely grown in Canada and *Brassica juncea* in India.

Canada, China and India are the world's largest producers (Table 3.4), with production in Canada benefiting from the wide

Table 3.4   Oilseed rape production in 2012 in countries producing at least 0.5 million tonnes *per annum*.

Country	Production (million tonnes *per annum*)
Canada	15.4
China	14.0
India	6.8
France	5.5
Germany	4.8
Australia	3.4
United Kingdom	2.6
Poland	1.9
Ukraine	1.2
United States	1.1
Czech Republic	1.1
Russia	1.0
Belarus	0.7
Lithuania	0.6
Denmark	0.5

Source: US Department of Agriculture.

adoption of genetically modified, herbicide-tolerant varieties since the mid-1990s. Germany, France and the UK are now also major producers, with Australia sandwiched between France and the UK. Yield is relatively low, at less than 4 tonnes of seed per hectare even in Europe; yield of oil is around one third of that, giving approximately 1.3 tonnes (1460 litres) per hectare, so higher than soybean but still much lower than palm oil (Section 3.4.3). Oilseed rape has only been a recognised crop for a relatively short period of time, and has therefore undergone a much shorter period of selective breeding; so there may be scope for yield improvement that has not yet been exploited.

Oilseed rape was used in the UK to provide oil for lubricants in the 19th century, and to some extent as a forage crop. The first time that it was grown widely was during the Second World War, when it was used to produce industrial oil. Its use as a forage crop was limited, and for human consumption almost nil, because the fatty acid that made up 50% of its oil was erucic acid (22:0); an oil that has similar properties to mineral oils but is toxic. Traditional industrial uses of rapeseed oil include the production of transmission oils, lubricants, oil paints; photographic film and paper emulsions, and healthcare products and plastics (in the form of erucic acid's derivative, erucamide).

Oilseed rape meal used to contain high levels of other undesirable compounds called glucosinolates. These are nitrogen- and sulphur-containing organic compounds that give mustard its 'hot' flavour and pungency. In large amounts they are bitter and toxic, causing vomiting and potentially even death. They are probably produced by the plant to deter herbivory, and may explain why oilseed rape and related species have relatively few natural pests.

Plant breeding over the thirty years after the Second World War reduced the levels of erucic acid and glucosinolates to the point where the oil from some varieties was considered acceptable for human consumption, and the first low erucic acid, low glucosinolate varieties (so-called 'double-lows') were grown in Canada in 1968. Nevertheless, oilseed rape did not get its seal of approval for human consumption (Generally Recognized as Safe) from the Food and Drug Administration of the US until 1985. Canadian producers then devised the name canola (Canadian oil low erucic acid) for edible oilseed rape oil. This name was adopted all over North America not only for the edible oil but also for the crop itself.

The oil of modern oilseed rape varieties for human and animal consumption is made up of oleic acid (18:1) (60%), linoleic acid (18:2) (20%) and α-linolenic acid (18:3) (10%), with palmitic, stearic and other saturated fatty acids together accounting for the other 10%. High erucic acid varieties are still grown today for industrial purposes but are not permitted to be placed in the food chain. However, the market is dominated by low erucic acid varieties, and most rapeseed oil biodiesel is therefore made from low erucic acid oil. It will be interesting to see if more high-erucic acid varieties are grown to satisfy the rapidly increasing demand for biodiesel, but low erucic acid rapeseed oil is regarded as a good feedstock for biodiesel anyway, with a relatively low cloud point of −10 °C and a cetane value typically of 51–55. It also has the advantages of not having to be kept separate from oil destined for the food chain, and of having a valuable animal feed co-product.

Oilseed rape is an annual crop and has a requirement for substantial applications of nitrogen and sulphur fertiliser (typically 190 kg nitrogen and 30 kg sulphur per hectare on sandy soils in the UK, for example). Nevertheless, the UK's Department of Transport calculates a conservative carbon intensity estimate for biodiesel from rapeseed oil used in the UK, ranging from 45 g/MJ if it originated in Poland, to 63 g/MJ if it originated in Australia; compared with 86 g/MJ for petroleum-based diesel. This gives a greenhouse gas saving of between 27 and 48%.

Rapeseed oil production in the EU, China and Canada is shown graphically in Figure 3.9. As with soybean, production has increased dramatically since the turn of the century, largely because of the oil's use in biodiesel, particularly in Europe. However, the

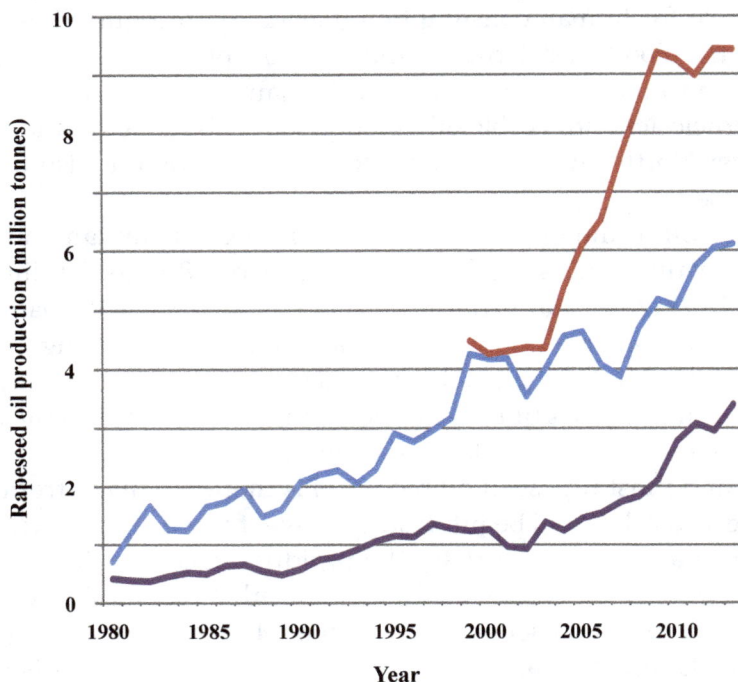

Figure 3.9   Rapeseed oil production from 1980 to 2013 in the EU-27 (red, from 1999 onwards), China (blue) and Canada (purple) (Data from the US Department of Agriculture).

increasing demand means that the price has also risen (Figure 3.10); from under US$400 per tonne in 2001 to peaks of over US$1700 per tonne in 2008, and US$1400 per tonne in 2011. At the end of 2013 it was just under US$1000 per tonne.

3.4.3 *Palm oil*

Indonesia and Malaysia have well-established industries based on the oil palm tree (*Elaeis guineensis*). Indeed, palm oil accounts for approximately half of global plant oil production, and almost 90% of the world's palm oil is produced in these two countries. Oil palm is a perennial crop that requires hot and humid conditions. It

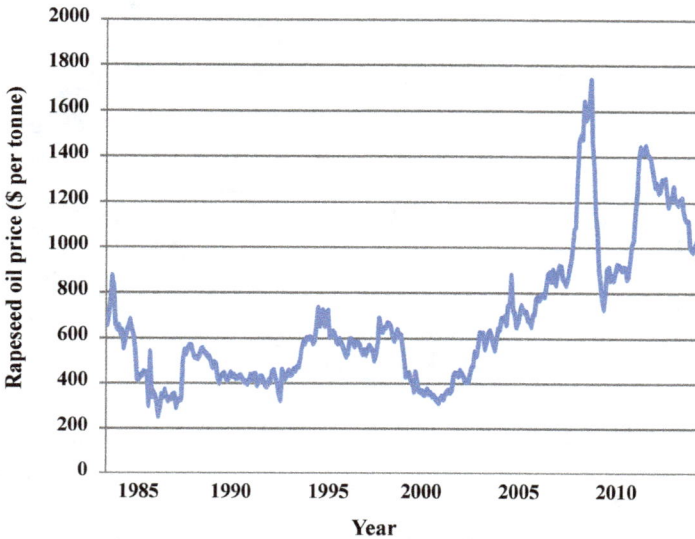

Figure 3.10   Rapeseed oil price from 1983 to 2013 (Data from World Bank).

originated in West Africa, but has been found to be adaptable for cultivation throughout the tropics. The trees take approximately 25 years to mature, at which point the yield of oil is typically around 4.2 tonnes (4700 litres) per hectare per year; more than ten times that of soybean oil. Oil is extracted from the fruits, which are each about the size of a plum, and are produced in large bunches weighing up to 40 kg each. Oil is extracted from both the flesh of the fruit and the kernel inside. Traditionally, these have been separated into palm oil and palm kernel oil, respectively; with the palm oil being used for edible products such as cooking oil, confectionery, margarines and other processed foods, and palm kernel oil being used in the manufacture of soaps, detergents and shampoos. Both palm oil and palm kernel oil are being used for biodiesel production and the distinction will not be made here.

Palm oil consists of the saturated fats palmitic acid (16:0) (45% of the total) and stearic acid (18:0) (5%); the monounsaturate oleic acid (18:1, omega-9) (40%), and the polyunsaturate linoleic

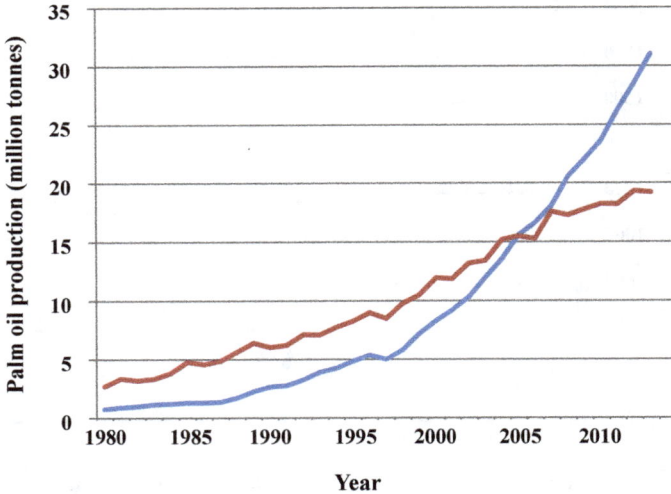

Figure 3.11   **Indonesia (blue) and Malaysia (red) palm oil production, 1980 to 2013 (Data from UN Food and Agriculture Organisation).**

acid (18:2, omega-6) (10%). The high proportion of saturated fatty acid makes its melting temperature relatively high, and it is semi-solid well above freezing. This means that biodiesel from palm oil has to be blended with other biodiesels or mixed with additives. Production of palm oil in Indonesia has almost quadrupled this century, from 8.3 million tonnes in 2000 to 31 million tonnes in 2013 (Figure 3.11), while that in Malaysia has increased from 11.9 million tonnes in 2000 to 19.2 million tonnes in 2013. About three quarters of the oil is exported, either as crude oil or after processing into cooking oil, and this is a major revenue generator for both countries. Both countries are also developing their own biodiesel industries, with Indonesia currently producing 1.26 billion litres per year and Malaysia 715 million litres per year.

As with other oils, there has been an increase in demand for palm oil for food uses, but the massive increase in production this century has occurred in part at least in response to the demand for biodiesel. This demand has led to a concomitant rise in the price of palm oil (Figure 3.12), from under US$200 per tonne in 2001 to a peak of over US$1200 per tonne in 2011, and a current price of

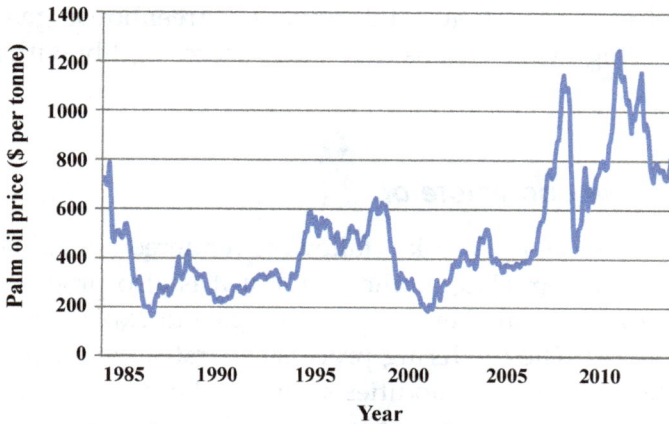

Figure 3.12 **Palm oil price from 1984 to 2013 (Data from World Bank).**

around US$800 per tonne; still four times the 2001 price, despite the increase in production that has been achieved over the same period. The Indonesian government clearly believes that this trend will continue, and has set a target of increasing the area of oil palm cultivation (currently 7.8 million hectares, of which 6 million hectares are being harvested) by an additional 4 million hectares.

This increase is not without its controversy. Despite its large population, Indonesia has a huge area of wilderness, much of it covered in rainforest. Indeed, the Indonesian rainforest is the third largest on Earth, behind those of the Amazon and the Congo, and the world's second most diverse habitat. Some of the increase in land area being used for oil palm cultivation may be converted from other agricultural uses, such as rubber and cocoa plantations, but it seems inevitable that some of it will come from what is now rainforest. Once again, the arguments for and against this process are not simple: as a perennial tree, oil palm is highly productive and requires relatively little input in terms of fertiliser or management. Its production costs are therefore very low; indeed biodiesel production from palm oil may be the only biofuel industry that could compete with petroleum-based fuel without political intervention. Its low input requirement also means that it has a low carbon intensity, calculated at 38 g/MJ even when imported and used in the UK (Table 1.3), giving a greenhouse gas saving of 56%. Its high yield means that it also scores highly with respect to land use.

3.4.4 *Tallow and waste oil*

Another biodiesel feedstock is tallow, or rendered animal fat. Most tallow comes from slaughterhouses, but it is also produced from restaurant grease, butcher shop trimmings and expired meat from grocery stores. The rendering process separates the fat from bone and protein, giving commodities such as yellow and white grease, as well as a high-protein by-product. This by-product used to be incorporated into some animal feeds but there have been strict restrictions on that since the bovine spongiform encephalopathy epidemic in the UK in the 1980s.

One problem with biodiesel produced from tallow is its low melting point: it starts to gel at around 16°C. Another is that it contains a relatively high concentration of free fatty acids as opposed to fatty acids incorporated into triacylglycerols. This means that biodiesel manufacture from tallow requires a pre-treatment with sulphuric acid and methanol, during which the free fatty acids are esterified with the methanol, followed by acid neutralisation and the alkali-catalysed transesterification of the triacylglycerol fatty acids. Consequently, tallow cannot simply be added to plant-derived biodiesel feedstocks. Nevertheless, tallow is a relatively cheap feedstock and an abundant product looking for a use.

Another product looking for a use is waste cooking oil and it has been suggested that this could also be used for biodiesel production, providing a cheap feedstock and getting rid of an abundant waste product that is difficult to dispose of cleanly. However, the unpredictability of the composition of the oil and the extent to which it may have become hydrogenated or oxygenated during storage and cooking may make this impractical.

3.4.5 *Other potential first generation feedstocks*

While oilseed rape is the major oilseed of temperate Europe, sunflower (*Helianthus annuus*) and olive (*Olea europaea*) are more suitable for cultivation in the warmer, drier climates of southern Europe. Sunflower oil typically comprises 7% palmitic acid (16:0), 4% stearic acid (18:0), 29% oleic acid (18:1) and 60% linoleic acid (18:2), although there are high-oleic varieties with oil containing 83% oleic acid and very little linoleic acid. The name oleic acid derives from *Olea*, so it is no surprise that there is a lot of oleic acid in olive oil, which typically contains around 70% oleic acid but in some varieties up to 83%. The other major constituents of olive oil are linoleic acid (18:2) and palmitic acid.

The EU is by far the biggest producer of both these oils, making 2.9 million tonnes of sunflower oil in 2013 and 2.5 million tonnes of olive oil, compared with 9.4 million tonnes of rapeseed oil. However, their value as food oils makes them significantly more expensive than rapeseed oil, the major European biodiesel feedstock; with sunflower oil trading in 2013 at over US$1100 per tonne, about 10% above the price of rapeseed oil (data from World Bank), and extra virgin olive oil trading at over US$3000 per tonne. This has limited the use of sunflower oil in European biodiesel production and precluded the use of olive oil altogether.

A relatively high price has also limited the use of coconut (*Cocos nucifera*) and peanut (*Arachis hypogaea*, also known as groundnut) oil in tropical and sub-tropical countries, with both trading at well over US$1000 per tonne and therefore uncompetitive with palm oil. Coconut oil also has an unusual composition, with 91% of it made up of relatively short-chain, saturated fatty acids: caprylic acid (8:0), capric acid (10:0), lauric acid (12:0) and myristic acid (14:0), as well as palmitic acid (16:0). The high saturated fat content means that it has good storage properties but a very high gelling temperature: 22°C for the oil and 10°C for the biodiesel produced from it. However, the Philippines, which is the world's largest producer of coconuts, is testing out a 5% blend of coconut biodiesel with petroleum-derived diesel in fuel for its public bus fleet.

3.5 Potential Second Generation Biodiesel Feedstocks

There has been a lot of interest in developing second generation (non-food) biodiesel crops that do not compete with food production. India has been at the forefront of research in this direction, because it is a highly populous country where the food versus fuel debate is particularly sensitive. It also has large areas of land (approximately 60 million hectares) that are classified as wasteland; that is, land where it is too arid and/or the soil is too poor to cultivate food crops. India is also the world's sixth largest consumer of energy, and diesel is the country's main liquid fuel; partly because of the country's extensive train network and large fleet of diesel locomotives. Currently, 80% of India's diesel is made from imported petroleum, and demand is predicted to rise steeply as the Indian economy grows. Consequently, in 2008 the Indian Government announced its 'National Biofuel Policy', with the aim of meeting 20% of the country's diesel demand with biodiesel. Approximately 40 million hectares of wasteland were selected for biodiesel production and the species that was chosen as the best option for producing the feedstock from this land was *Jatropha*.

Jatropha (*Jatropha curcas*), or physic nut, is a type of flowering shrub of the *Euphorbiaceae* family that actually comprises at least 170 different species worldwide. The 'physic' part of the common name is actually a shortening of physician, reflecting the use of the oil in traditional medicine. Any medical use involves very small quantities because the nut and oil are extremely toxic due to the presence of a multitude of hazardous compounds, including a highly poisonous lectin-type protein called curcin. For this reason the oil is sometimes called vomit or purge oil (this is definitely not a food crop).

The presence of curcin and other toxic compounds makes *Jatropha* highly resistant to pests and it is also very tolerant of drought and poor soil quality, making it suitable for low-input cultivation on the arid wastelands that India wants to develop for biodiesel production. Under prolonged drought conditions the plant survives by shedding its leaves. Its seeds consist of up to 40% oil, the composition of which may vary between species but is typically around 15% palmitic acid (16:0), 6% stearic acid (18:0) (making the saturated fat content approximately 21%), 42% oleic acid (18:1) and 35% linoleic acid (18:2), with minor constituents making up the other 2%. The yield of oil is around 1.4 tonnes (1600 litres) per hectare, which is comparable with oilseed rape oil but still some way below palm oil. *Jatropha* biodiesel has a cetane value of 46; its low temperature performance is not particularly good (its cloud point is 2 °C) but that is not a big issue for most domestic Indian consumption.

Rural Indian communities extract the oil from *Jatropha* seeds using a simple screw press, and purify it by sedimentation or filtration. The unprocessed oil is often used to fuel the single-cylinder diesel engines that are typical of agricultural machinery and domestic generators in India. It is also blended with petroleum-derived diesel. There is now a growing industry to convert the oil to biodiesel by transesterification. However, the oil contains a relatively high proportion of free fatty acids (typically 14%), and, like tallow, has to be pre-treated with acid first.

The toxicity of *Jatropha* oil is a significant disadvantage because it means that, currently, there is little use for the meal that remains after the oil has been extracted. It may be possible in the future to use the meal as feedstock for second generation bioethanol production from the cellulose that it contains (Section 2.5), or for biogas production (Chapter 5), but at present it can be used for compost and not much else.

Another plant that survives in arid, low nutrient conditions is castor bean (*Ricinus communis*). Like *Jatropha* oil, castor bean oil has been used in traditional medicine and like *Jatropha* oil it is extremely toxic, with toxicity imparted by, amongst other things, another lectin-type protein called ricin. Castor bean yields up to 0.9 tonnes of oil per hectare, which is less than *Jatropha*, but this species may be more suitable than *Jatropha* in some areas and has been grown commercially to produce industrial oils for considerably longer, so there is more experience of its cultivation.

A third candidate for second generation biodiesel production in the tropics and sub-tropics is *Pongamia* (*Millettia pinnata*; previously *Pongamia pinnata*), a medium-sized leguminous tree in the *Fabaceae* (pea) family that produces large, oil-rich seeds. Like *Jatropha*, *Pongamia* tolerates arid conditions and can be grown on land too poor for food production. Its oil is toxic, but not as toxic as *Jatropha* oil, and the meal could possibly be incorporated into animal feed if mixed with other constituents. The big advantage that *Pongamia* has is the fact that it is a legume, and like other legumes it acquires nitrogen through a symbiotic relationship with soil bacteria, collectively called rhizobia, that fix atmospheric nitrogen. This means that it does not require nitrogen fertiliser.

Pongamia oil typically makes up about 40% of the seed and can be extracted with a simple mechanical press, which, as with *Jatropha*, is important for rural communities. It consists of around 54% oleic acid (18:1), 19% linoleic acid (18:2), 11% stearic acid (18:0), 7% palmitic acid (16) and 9% minor constituents. The cloud point of *Pongamia* biodiesel is typically over 8°C, so, like *Jatropha* biodiesel, it is suitable only for use in tropical and sub-tropical regions unless mixed with something else.

There is still a lot to learn about the large-scale cultivation of these species. For *Jatropha* and *Pongamia*, there is no experience at all of commercial use, and even castor bean has previously been no more than a small-scale niche crop. Data on yields, where there is any, comes from small-scale trials; and large-scale cultivation in arid wasteland, as is proposed, may not give anything like the same yield. Harvesting the seeds from *Jatropha* shrubs and *Pongamia* trees is still done by hand, with little sign of mechanisation being developed, and in *Jatropha* the seeds are produced sporadically throughout the year, making the planning of harvesting and processing difficult.

Another impediment to the development of trees like *Jatropha* and *Pongamia* for biodiesel is that there is no experience of breeding them, and hardly any knowledge of their genetics or the traits that could be bred for. Furthermore, the aim, at least, is that the trees will be grown on low quality land by relatively poor rural communities who would not have the resources to pay for genetically improved varieties, and once planted the trees would be productive for many years anyway, so there would be no need to replace them and therefore little opportunity for plant breeders to recover their investment. With these challenges to be overcome, it will be interesting to see if the oil can be produced in the quantities that are required and in an economically competitive manner.

Another possible second generation feedstock for biodiesel production is algae. The potential for using algae to produce bioethanol was discussed in Section 2.5.2, and the same arguments in favour of using algae also apply to biodiesel; the main ones being the fast growth rate and huge potential yield. Algae make and accumulate fatty acids just as higher plants do; indeed they make longer polyunsaturated fatty acids than those found in higher plants, including the omega-3, long-chain polyunsaturated fatty acids, eicosapentaenoic acid (EPA) (20:5) and docosahexaenoic acid (DHA) (22:6). EPA and DHA are required for foetal and neo-natal development, and both are commonly added to infant formula milk. They are present in fish oils, but fish do not synthesise them; rather, they are made by algae and accumulate through the marine food chain. Some algal species accumulate fatty acids up to 60% by weight, and yields of more than 100,000 litres per hectare have been suggested to be achievable from heterotrophic systems; more than 20 times the yield of palm oil, which is the most productive land-based system. Algal fatty acids can be converted into fatty acid methyl esters for biodiesel, just as fatty acids from land plants can.

A company at the forefront of developing algal oil production systems is Solazyme, based in California, US. Solazyme claims to be close to marketing a range of biodiesels based on fatty acid methyl esters of algal oils, including road vehicle and marine blends. It is also developing an algal-based jet fuel. The company's first commercial-scale production lines, incorporating 625000 litre fermentation tanks, went into operation in Brazil in May 2014, in a joint venture with Bunge Global Innovation. Production is expected to reach capacity in mid to late 2016. The company information on the development does not give details of the algal species involved.

As with bioethanol, the potential of algal-based systems appears huge, but the same caveat applies, in that there is no experience of using algae for commercial production on this scale. The success or failure of Solazyme's venture will therefore be closely watched by other potential investors.

3.6 Biotechnology

It is impossible to discuss any issue relating to soybean and oilseed rape production without considering biotechnology. According to the International Service for the Acquisition of Agri-Biotech Applications, 81% of global soybean production and 30% of oilseed rape production now involves genetically modified (GM) varieties, despite resistance to adoption of the technology in some areas of the world, notably Europe. GM varieties have been adopted in most of the major soybean exporting countries, including the US, Brazil and Argentina; and GM oilseed rape is popular in Canada.

Most of these GM varieties carry a trait that imparts tolerance to a broad-range herbicide, and this is not directly relevant to biodiesel production except that it reduces production costs. In the case of soybean the herbicide in question is glyphosate. This herbicide targets an enzyme called 5-enolpyruvoylshikimate 3-phosphate synthase, which is essential for the production of amino acids with aromatic side chains (phenylalanine, tyrosine and tryptophan). Tolerance is imparted by a bacterial gene that encodes a version of the enzyme that is not affected by the herbicide. Glyphosate is marketed by the agricultural biotech company Monsanto as 'Roundup', and glyphosate tolerance is the basis for its 'Roundup-Ready' soybeans, which have been on the market since 1996. Since 81% of soybeans worldwide are now GM herbicide-tolerant

(GM-HT), it follows that most of the soybean oil used for biodiesel, like anything else, comes from GM soybeans.

Glyphosate tolerance has now been engineered into oilseed rape as well, but in that species there is competition from another GM-HT trait that imparts tolerance to the herbicide gluphosinate. Gluphosinate tolerance technology, which is marketed by Bayer, is based on another bacterial gene that encodes an enzyme called phosphinothrycine acetyl transferase, which breaks the herbicide down. The herbicide is marketed under the trade name Liberty, and varieties carrying the tolerance trait have the trade name LibertyLink. Approximately 30% of global oilseed rape production is now based on GM-HT varieties, although they are currently not licensed for use in Europe and there is little prospect of them being licensed in the near future.

More directly relevant to the biodiesel industry is another GM soybean variety that is already on the market, and which has altered oil composition. In conventional soybean, oleic acid (18:1) is synthesised in the plastids and transported to the endoplasmic reticulum as oleoyl-ACP (Figure 3.3). It is converted to linoleic acid (18:2) in the endoplasmic reticulum by a single desaturation step carried out by an enzyme called a Δ12-desaturase, encoded by the gene *FAD2* (fatty acid desaturase-2). Palmitic acid (16:0) is also released from the plastid, and this process is controlled by an enzyme called palmitoyl-ACP thioesterase. More palmitoyl-ACP thioesterase activity results in more palmitoyl-ACP being released from the plastid and less being converted to stearoyl-ACP and oleoyl-ACP. Palmitoyl-ACP thioesterase is encoded by a gene called *FATB*. A GM soybean line was developed in the early 1990s with reduced *FAD2* gene expression and consequently less Δ12-desaturase activity. It accumulates oleic acid to approximately 85% of its total oil content, compared with approximately 20% in non-GM varieties. It also contains less palmitic acid. This variety was produced by PBI, a subsidiary of DuPont, and is marketed under the trade name 'Vistive'. A second modification that targets the *FATB* gene has since been developed, and oil from soybean with reduced expression of both *FATB* and *FAD2* is 95% oleic acid.

Monsanto have also produced a high oleic acid variety, marketed as 'Plenish', but while it carries a GM-HT trait imparting tolerance of glyphosate, the high oleic acid trait results from mutagenesis, not genetic modification.

The main reason for developing these varieties is that oleic acid is very stable during frying and cooking, and less prone to oxidation during storage than polyunsaturated fats, making the oil less likely to form compounds that affect flavour. The traditional method of preventing polyunsaturated fat oxidation involves hydrogenation, and this runs the risk of creating *trans* fatty acids. *Trans* fatty acids contain double bonds in a different orientation to the *cis* fatty acids present in natural plant oils. They behave like saturated fat in raising blood cholesterol, and potentially contributing to blockage of arteries. Indeed, although technically unsaturated, they cannot be described as such in US food labelling. The oil produced by high oleic acid GM soybean requires less hydrogenation, and consequently there is less risk of *trans* fatty acid formation. Hydrogenation also generates oils that are solid at ambient temperatures, and makes a biodiesel with poor low temperature performance, high viscosity, and low lubricity. High oleic acid soybean oil should therefore be a better feedstock for biodiesel production. The reduction in polyunsaturates should also improve the cetane number, and while oleic acid has a lower melting point than linoleic and linolenic acids, the effect of this is outweighed by the reduction in saturated fats.

Oilseed rape oil has also been the target for biotechnologists. The company Calgene, subsequently taken over by Monsanto, genetically modified an oilseed rape variety to produce high levels of lauric acid in its oil. This variety was introduced onto the market in 1995. It contained a gene from the Californian Bay plant that encodes an enzyme that causes premature chain-termination of growing fatty acid chains. The result is an accumulation of the 12-carbon chain lauric acid to approximately 40% of the total oil content, compared with 0.1% in unmodified oilseed rape, making the oil an alternative to palm and coconut oil as a source of fatty acids for the manufacture of detergents. However, as discussed above, palm oil production in particular is very efficient, and oil from the GM oilseed rape did not gain a foothold in the market. Cultivation of this variety was never more than small-scale and has now ceased. Nevertheless, this episode does show that oilseed rape oil can be altered using GM techniques and the development of the biodiesel industry may lead to new targets being identified in the future.

4 BIOMASS

4.1 Introduction

Electricity generation is currently heavily dependent on the burning of fossil fuels, in particular coal. The carbon intensity of energy generation from coal is the highest of all the fossil fuels at 112 g/MJ (Table 1.2), and it also produces sulphurous and nitrogenous pollutants that nowadays are subject to strict controls in most countries. World coal consumption annually exceeds 7 billion tonnes, and despite coal's pollution issues is predicted to increase to over 9 billion tonnes over the next two decades. Almost half of this consumption is in China, which generates more than two thirds of its electricity from coal. Western countries are also heavily dependent on coal, with the US, for example, generating just under half of its electricity from burning coal. Given coal's high carbon intensity, finding alternative sources of energy for electricity generation could potentially provide significant reductions in greenhouse gas emissions; and biomass, used either for co-firing with coal or on its own, could provide one such alternative.

Biomass is solid material from biological (mainly plant) sources. Charcoal could therefore be considered to be a form of biomass, and charcoal continues to be an important fuel in many parts of the world, with 1.87 billion cubic metres of roundwood being used for fuel, most of it after conversion to charcoal, in 2012. Indeed, there has been a resurgence in the use of charcoal for smelting in Brazil in an effort to reduce carbon emissions associated with the Brazilian steel industry. This charcoal is coming from traditional forestry, and the strategy will only contribute to a reduction in greenhouse gas emissions if the forestry is managed sustainably, because deforestation is a major contributor to greenhouse gas emissions and climate change. However, the term biomass is not usually applied to traditional charcoal, but reserved for the use of sustainable waste products and material from novel feedstocks such as willow, poplar, *Miscanthus*, other grasses and reeds.

In the early years of bioenergy development and thinking, the generation of electricity from novel sources of biomass was expected to be the first form of bioenergy to become a commercial reality. It turned out that liquid biofuel for transport developed much faster; and while that industry is now well established, as described in the two previous chapters, electricity production from biomass remains relatively small-scale, and in the UK, at least, mainly uses waste material from traditional forestry rather than novel crops for feedstock. The small-scale nature of electricity production from biomass may be about to change, but willow, poplar, *Miscanthus*, other grasses and reeds can all be considered as second generation biomass crops. They are described in detail later in the chapter.

In the UK, the use of biomass for electricity production is being ramped up in response to the country's Renewables Obligation Schemes, which are described in Chapter 1, and are requiring electricity suppliers to generate an increasing proportion of their electricity from renewable sources. Wind, hydroelectric, solar and tidal energy all qualify as renewable, and there has been a huge growth in electricity from wind and solar energy in the past decade. However, as elsewhere, most of the UK's electricity is still generated from power stations that burn fossil fuels. Indeed, the largest power station in Western Europe is the Drax Power Station in North Yorkshire, which generates almost 4000 megawatts and provides approximately 7% of the UK's electricity supply. Until recently the power station burnt coal, and although flue gas desulphurisation equipment has been fitted and the company running the power station (Drax Group plc) claims it is the cleanest coal-fired power station in the country, Drax is the UK's largest emitter of CO_2.

Drax started co-firing biomass along with coal in 2002, in order to comply with the Renewables Obligation Scheme. In 2012 it announced plans to go much further; initially applying for planning permission to build dedicated biomass-firing generators on the site, then changing its plans to convert three of its six coal-fired generators. The first of these became operational in 2013 and the third is scheduled for 2017, at which point the largest power station in Western Europe will be a predominantly biomass-fuelled operation, an astounding development.

When fully operational, the three biomass-fired generators will burn 7.5 million tonnes of biomass per year. Currently the biomass is almost entirely imported from the US and Canada and is a by-product of traditional forestry; that is sub-standard wood from diseased and damaged trees, small branches and trimmings etc. The US Department of Energy calculates that US forestries produce 93 million tonnes of such material annually. The material has to be transported to the UK by ship, of course, in the form of wood pellets. Even so, Drax claims an 80% reduction in CO_2 emissions compared with burning coal, giving the material a carbon intensity of $23 \, g/MJ$. The company is also inviting would-be suppliers to contact them, meaning that there is a large potential market for biomass produced in the UK and Europe if it can be supplied competitively. This market could see the development of biomass as a valuable co-product for established forestries, and/or encourage the cultivation of second generation biomass crops.

Drax claims that it only burns sustainable biomass, and there seems little reason to doubt this because forestry in the US and Canada is a highly regulated and monitored industry. The company has adopted a policy that is consistent with the framework set out by the Nuffield Council on Bioethics, claiming that it gives preference to sources that maximise greenhouse gas savings, do not result in a net release of carbon from vegetation or soil, do not endanger food supply or communities where the use of biomass is essential for subsistence and do not adversely affect protected or vulnerable biodiversity, while preserving and protecting soil and water, and contributing to local prosperity and social well-being. Assuming it stands up to scrutiny, all this seems very laudable and it will be interesting to see what effect Drax's transformation has on its standing with environmental campaign groups, who previously are more likely to have seen Drax as part of the climate change problem rather than part of the solution.

Drax has identified a source of biomass that more than meets its needs. However, even the 93 million tonnes of forestry waste that the US produces each year is dwarfed by the world's annual 7 billion tonne coal consumption, and if other electricity generators follow Drax's lead, demand may soon outstrip supply, even if the global forestry industry becomes involved. Other agricultural waste products such as straw, sugar cane and beet bagasse (Sections 2.3.2 and 2.3.3) may be used, but markets may also open up for biomass from second generation biomass crops, and these are the subject of the rest of this chapter.

4.2 Willow

4.2.1 *The suitability of willow as a biomass crop*

Willows are deciduous trees and shrubs of the genus *Salix*. There are at least 400 species in the genus, but a few are known by the common names sallow or osier rather than willow. All are native to the temperate latitudes of the northern hemisphere. The name of the genus is derived from the Celtic words *sal* meaning water and *lis* meaning near. Willows are often found near water because their seeds require a plentiful supply of moisture for germination. A derivative of the genus name, salicin, is the name given to a chemical (formula $C_{13}H_{18}O_7$) found in the bark. Salicin can be converted to salicylic acid (salicylate), a hormone that is involved in invoking and co-ordinating the plant's defence response against herbivores and pathogens. Salicylic acid also has anti-inflammatory, anti-fever and analgaesic properties; indeed aspirin (acetylsalicylic acid) is a pro-drug for salicylic acid (i.e. it is converted to salicylic acid when ingested).

The presence of salicin probably explains the use of willow bark in herbal medicine. Traditionally, willow has also been used to make baskets, fences and cricket bats, for which its strength, flexibility and light weight make it particularly suitable. It has also been grown to produce charcoal, so it has some history as an energy crop. Willow has several characteristics that make it a suitable potential biomass energy crop:

- Fast growth and coppicing ability
  Fast growth is clearly an important characteristic of a potential biomass crop, and in willow the natural capacity for rapid accummulation of biomass is augmented by cultivation as short rotation coppice. This practice has been used for centuries and involves allowing the tree to establish itself then cutting the stem back during winter dormancy to just above the roots to leave a stool (a stump from which new shoots will grow). The following spring the stool throws out multiple shoots from axillary buds that were previously kept dormant by hormones produced by the main shoot (apical dominance) (Figure 4.1). There follows a period of several years of rapid growth of these multiple shoots to give a tree like those shown in Figure 4.2, at which point the stems are harvested back down to the stool. This process can be repeated many times. The stems are dried and chipped for distribution to energy generators. Yields of up to 20 tonnes dry weight per hectare per year have been obtained from field trials in the UK and over 10 tonnes per hectare in some commercial plantations.

- Low requirement for nitrogen fertiliser
  Willow is, of course, a perennial plant, and while there is a requirement for nitrogen fertiliser to replace what is lost at harvest, the fact that the plant does not have to regrow from a seed means that it needs much less nitrogen than annual crops. This reduces costs and carbon intensity.

Figure 4.1   A coppiced willow tree with multiple stems emerging from the stool (picture courtesy of Rothamsted Research Visual Communication Unit).

- Ability to grow on marginal land
  As with the second generation biodiesel species, *Jatropha* and *Pongamia*, an attraction of willow as a biomass crop is that it can grow on land that is not considered suitable for arable crop production, meaning that it could be grown without affecting food supply.

- Ease of propagation and cultivation
  Willows will grow readily from stem segments, usually of 15–20 cm in length, simply pushed into the soil, which makes establishment of a plantation relatively easy. Planting is undertaken in spring and the trees are coppiced at the end of the first growing season. Weed control may need to be applied in the first year, but after that the shading from the trees is sufficient to prevent significant competition from weeds. In the UK, the trees are harvested every three years, and willow plantations remain productive for around 25 years before they need to be replanted.

Figure 4.2   Willows growing in short rotation coppice at Rothamsted Research in the United Kingdom (picture courtesy of Rothamsted Research Visual Communication Unit).

- Potential for further improvement
  As with other second generation bienergy crops, the adoption of willow for commercial cultivation may be hampered by lack of familiarity and confidence amongst farmers, and of experience in its cultivation. Farmers may also be reluctant to take the risk of investing in a crop that will not generate revenue for

several years. The flip side of commercial production of a species that has not been cultivated intensively before is that there may be considerable potential for genetic improvement that has not yet been exploited. In the case of willow, there is a wealth of genetic diversity that could be brought into breeding programmes, in the form of the hundreds of different related species, subspecies and ecotypes within the genus from a wide range of environments. Most of these species hybridise readily.

- Winter harvesting
  Willow is harvested in winter and there are two advantages of this. Firstly, it fits well with other farm operations. Secondly, the plant withdraws most of its nutrients down to the roots in preparation for winter dormancy, so the harvested stems contain relatively little nitrogenous and sulphurous compounds that would generate pollutants on burning.

4.2.2 *Current commercial cultivation of willow*

As described at the beginning of this chapter, there is now an emerging market for biomass in the UK as Drax power station converts parts of its operation to burn biomass instead of coal. However, although Drax's early experimentation with biomass as a fuel was with willow, this market is currently being supplied by waste forestry product from the US and Canada. Dedicated biomass crops such as willow have not yet been adopted as commercial crops on a large scale and, although there is some cultivation of willow, it is as a niche crop, grown in some cases by unexpected landowners. East Midlands Airport, in Leicestershire, for example, has a small willow plantation as part of its commitment to achieve 'carbon neutral' operations (it has also installed two wind turbines). The willow plantation is only of 26 hectares, so it may help the airport with its carbon figures but it is hardly going to supply a commercial power station. It will be interesting to see if Drax's creation of a large, reliable market for biomass encourages more farmers and foresters in the UK to grow willow.

Larger-scale cultivation of willow as an energy crop is already established in Sweden, where it has become an important contributor to energy generation over the last 20 years. There are now more than 16,000 hectares of willow plantations in the country, making Sweden the largest willow biomass producer in Europe. There is a high level of mechanisation, from planting to harvesting, and the stems are usually chipped on site before transport to customers. However, the harvest interval is slightly longer than in the UK, typically every three to five years, and yields are typically only five to six tonnes per hectare per year.

4.3 Poplar

Poplar (*Populus*) is another diverse genus of trees comprising hundreds of species, including not only those commonly called poplars but also aspens and cottonwoods. Poplars are related to willows, both being members of the *Salicaceae* family, and, like willows, occur in a range of mainly temperate habitats with a broad geographical distribution. Unlike willows, however, they are already cultivated extensively in modern forestry operations, being used for paper, veneer and lumber production. Forestry is based on three sub-types: Aigeiros, which includes black poplar (*Populus nigra*) and eastern cottonwood (*Populus deltoides*); Tacamahaca, which includes the balsam poplars; and Populus (formerly called Leuce), which includes the aspens and white poplar (*Populus alba*). Currently just over 5 million hectares of poplar are cultivated worldwide (data from International Poplar Commission of the United Nations), most of it (4.3 million hectares) in China and almost all of it for traditional forestry uses.

Poplar has many of the attributes associated with willow that make it suitable for biomass production, in particular a high growth rate, amenability to short rotation coppicing, low nitrogen demand, easy propagation from stem pieces and wide genetic diversity available for breeding programmes. As an established forestry species there is also more experience of its cultivation, although not necessarily in the short rotation coppice systems designed for biomass production. Poplar has also become something of a model tree for genetic research thanks to its relatively small genome size. Indeed, the nucleotide sequence of the entire poplar genome has been in the public domain for several years, and it was one of the first tree species for which genetic modification was demonstrated; something that is still not routine for willow. Its disadvantages include high water demand and a reputation for susceptibility to disease and insect herbivory.

Poplar produces fewer but thicker stems than willow when coppiced, and has a longer rotation time; typically four years in the UK rather than three. So far, willow appears to have been preferred for biomass production, although there are experimental poplar plantations in the US, UK, Italy, Eastern Europe and Argentina. Yields in these experimental trials are typically 10–15 tonnes per hectare per year, although yields exceeding 20 tonnes per hectare per year have been reported.

4.4 *Miscanthus*

Miscanthus (Figure 4.3) is a genus of perennial grasses native to subtropical, tropical and temperate regions of south and east Asia. *Miscanthus* species are characterised by rapid growth, low mineral content and high yield of biomass; characteristics that have led to its selection as a potential bioenergy crop. It has mainly been considered for biomass production, with biomass being dried and pelleted after harvest. However, there is also interest in it as a source of cellulose for bioethanol production (Section 2.5). *Miscanthus* is sometimes called 'elephant grass' but it is not the same species as African elephant grass (*Pennisetum purpureum*).

Miscanthus is a member of the *Andropogoneae* tribe, which also includes maize, sorghum and sugarcane. The genus comprises at least 14 species and probably more, the taxonomy being made more complicated by the existence of many inter-species hybrids. The species include *Miscanthus sinensis*, which is distributed from Hebei province in northern China to Hong Kong, Korea, Taiwan and Japan, *Miscanthus sacchariflorus*, which ranges from Russia

Figure 4.3 *Miscanthus* growing in field trials at Rothamsted Research in the United Kingdom (picture courtesy of Rothamsted Research Visuual Communication Unit).

through northern China to Korea and Japan, *Miscanthus floridulus*, which inhabits sub-tropical and tropical south-east Asia, and *Miscanthus tinctorius*, from which a traditional yellow dye is extracted in Japan.

There are both diploid and tetraploid forms of *Miscanthus* and the two forms will hybridise to give fast-growing, vigorous, triploid hybrids. One of these triploids, *Miscanthus giganteus*, is a sterile hybrid of *Miscanthus sinensis* and *Miscanthus sacchariflorus*. *Miscanthus giganteus* has been a favourite of field trials in Europe since the early 1980s. It reaches heights of more than 3.5 metres in one season and can yield up to 25 tonnes of biomass per hectare, although yield even in field trials is highly location-dependent.

One reason why *Miscanthus* can produce so much biomass in one season is that it will over-winter as a rhizome (underground stem), from which new shoots grow in spring, with the rhizome providing nutrients for the new growth. Nutrients are then withdrawn back into the rhizome as the above-ground parts of the plant senesce in the autumn. Harvesting takes place in early winter, so the amount of nitrogenous and sulphurous compounds that would form pollutants on burning is reduced. There are, therefore, some similarities with short rotation coppiced willow, in both the rapid production of new biomass from a perennial base and the low concentration of compounds that will form pollutants. Another similarity with willow that ticks a box for a potential energy crop is a low demand for nitrogen fertiliser.

Parameters on which *Miscanthus* scores poorly include frost tolerance, with *Miscanthus giganteus* rhizome survival rates declining when the soil temperature falls below 3°C. New plantations are also slow to establish in European climates and require weed control until the canopy develops. As a sterile hybrid, *Miscanthus giganteus* produces no seed and propagation relies on cloning, which significantly increases the cost of establishing a plantation. The development of fertile genotypes that could match *Miscanthus giganteus* yield would address this problem but would raise the possibility of the species becoming an invasive weed in areas for which it were particularly well-adapted, since it is not a native species.

The Biomass Crop Assistance Programme run by the United States Department of Agriculture has supported the development of commercial plantations of *Miscanthus giganteus*. Aloterra Energy and MFA Oil Biomass, for example, based in Arkansas, have used this federal support to plant over 7,000 hectares of *Miscanthus giganteus* at four different sites, citing high yields, low nutrient requirements, non-invasive characteristics and adaptability to a wide growing area as reasons for opting for *Miscanthus giganteus* in favour of other biomass crops.

4.5 Native American Prairie Grasses

The natural ecosystem of the Great Plains of North America between the Rocky Mountains and the Mississippi River is

Figure 4.4 Switchgrass growing in field trials at Rothamsted Research in the United Kingdom (picture courtesy of Rothamsted Research Visual Communication Unit).

tall-grass prairie, and several of the native species of fast-growing, perennial prairie grasses are being investigated as potential biomass crops. One of the dominant species of the prairies is switchgrass (*Panicum virgatum*) (Figure 4.4), which ranges from southern Canada as far south as Mexico. Like *Miscanthus*, switchgrass grows rapidly in spring from an over-wintering rhizome and it can reach almost 3 metres in height by the end of the season. It can grow on marginal lands with low inputs and is considerably hardier than *Miscanthus*.

Another candidate is big bluestem (*Andropogon gerardii*), also known as tall bluestem, bluejoint or turkeyfoot. The name big bluestem derives from the fact that the stem turns blue or purple at maturity. Big bluestem also grows rapidly from an over-wintering rhizome and, like switchgrass, tolerates poor soil conditions. It grows to a similar height as switchgrass at 3 metres. A third species attracting interest is indiangrass (*Sorghastrum nutans*), also known as yellow indiangrass.

4.6 Giant Reed and Reed Canary Grass

Giant reed (*Arundo donax*), also known as giant cane, wild cane, Spanish cane, carrizo or arundo is a tall, perennial cane reed that grows in damp soils near water sources. Its natural range extends from the Mediterranean region to India and Nepal, but it has also been planted in sub-tropical and tropical regions of the US and the Carribean. Giant reed forms dense stands up to 10 metres in height, with thick, hollow stems. It flowers in late summer but forms infertile seeds or no seeds at all, reproducing vegetatively from tough, vigorous, underground rhizomes. New plants can be generated readily from short rhizome or stem pieces.

Another species that prefers damp conditions is reed canary grass (*Phalaris arundinacea*) (Figure 4.5). This species grows along

Figure 4.5   Reed canary grass (foreground) growing in field trials at Rothamsted Research in the United Kingdom (picture courtesy of Rothamsted Research Visual Communication Unit).

the margins of lakes over a huge range, from Europe to Asia, -northern Africa and North America. It is another perennial grass that forms thick, underground rhizomes, and the stems can grow rapidly from the rhizome to a height of 2 metres.

4.7 Co-Firing of Grain

There are anecdotal reports that some electricity generators in the UK have attempted to meet their targets for using renewable sources of fuel by co-firing cheap, low-quality wheat grain. Wheat and other cereal grain has a high protein content, and the nitrogen from the protein will be given off in the form of nitrous oxide (N_2O), nitric oxide (nitrogen monoxide (NO)) and nitrogen dioxide (NO_2) during burning. As stated at the beginning of this chapter, CO_2 is the most important greenhouse gas, not because it is the most effective at preventing thermal energy from escaping the Earth but because far more of it is being emitted as a result of human action than anything else. Nitrous oxide (also known as laughing gas because of the euphoric effects of inhaling it) is actually a far worse greenhouse gas, with 298 times the global warming potential of CO_2. Nitrogen dioxide, which is a heavy, brown gas, is also an important air pollutant because it is toxic by inhalation. Clearly, burning grain as biomass makes absolutely no sense, and is an unacceptable practice.

4.8 Thermal Conversion

While direct burning is the most obvious and commonly used method of generating energy from biomass, the material can first be subjected to thermal conversion to make derivatives that have superior combustion characteristics. Thermal conversion involves high temperatures in the complete absence of oxygen (pyrolysis) or the presence of limited amounts of oxygen that result in partial oxidation (gasification). Indeed, as discussed in the Introduction to this chapter, traditional charcoal production involves pyrolysis of wood at temperatures up to 800°C. The difference with the modern approach is that the gases that are produced are trapped and condensed to form bio-oil, a mixture of hydrocarbons that can be used as a heating oil or fractionated to make petrol and diesel.

Gasification also requires temperatures of around 800°C, but the presence of some oxygen, which can be supplied in the form of pure oxygen, air or steam. The biomass initially forms charcoal and a mixture of CO_2 and water vapour. The CO_2 and water vapour are reduced by the charcoal to carbon monoxide (CO) and hydrogen. Gasification with air produces a low value heating gas, the nitrogen from the air reducing its calorific value; but gasification with oxygen or steam produces a medium value heating gas and 'syngas', a mixture of carbon monoxide (40%), hydrogen (40%) methane (3%) and CO_2 (17%). Syngas can be used as a feedstock for the manufacture of methanol, ammonia, synthetic diesel or substitute (synthetic) natural gas (SNG).

5 BIOGAS

5.1 Introduction

Biogas is a mixture of methane (CH_4) (Figure 1.4) and carbon dioxide produced by the anaerobic digestion of biological material. It can be used as a fuel itself, or the methane can be separated from the carbon dioxide to produce biomethane (sometimes called renewable natural gas, or RNG).

Anaerobic digestion, that is the decomposition of biological material in the absence of oxygen, is a natural process and is responsible, for example, for the production of marsh gas, which can form over bogs, and for the methane that is produced in the guts of ruminant animals such as cattle and sheep. It is also an important part of sewage and waste-water treatment, and methane makes up a proportion of sewage gas. The process also goes on quite naturally in garden composters, piles of farmyard manure and silage, and landfill sites.

The process of anaerobic digestion is often attributed to bacteria but, in fact, although bacteria are involved in the process, it is micro-organisms of the Archaea kingdom that are actually responsible for the production of methane. These micro-organisms used to be classed as bacteria, but are now recognised as having a number of fundamental differences from bacteria and have been re-classified as a completely separate kingdom. The Archaea that are responsible for methane production are called methanogens and more than 50 species have been identified. The different species are not all closely related, so the term methanogen simply denotes the fact that they produce methane and does not represent a formal evolutionary classification. They are strictly anaerobic and most will die in the presence of even trace levels of oxygen.

5.2 Anaerobic Digestion

A flowchart for the anaerobic digestion process is shown in Figure 5.1. The process has four stages:

- Hydrolysis
- Acidogenesis
- Acetogenesis
- Methanogenesis

The first stage, hydrolysis, involves the breakdown of insoluble lignocellulose, hemi-cellulose, cellulose (Section 2.5), oils, proteins

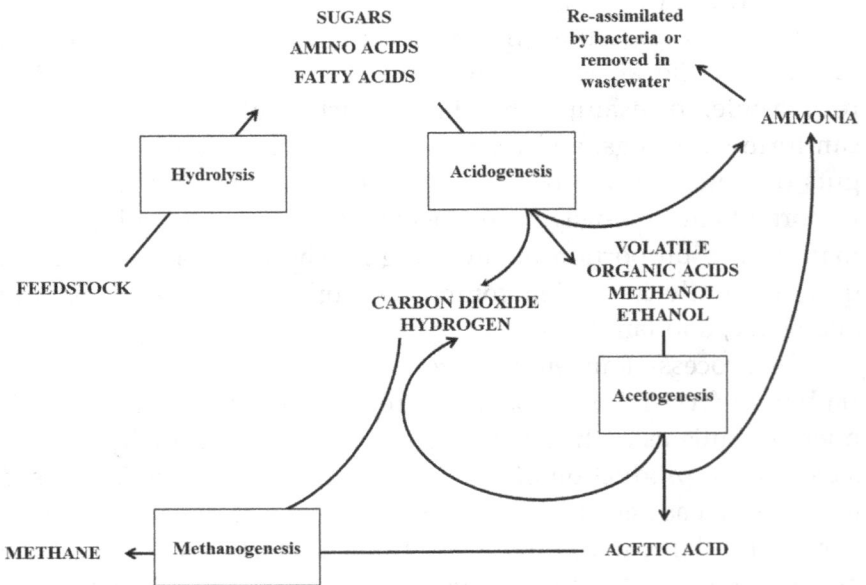

Figure 5.1　Flowchart for the anaerobic digestion process.

and other complex polymers by bacteria. Even if the input material is sewage or manure, lignocellulose, hemi-cellulose and cellulose (dietary fibre) make up much of the content because of their indigestibility. They also make up a large proportion of food waste, partly because meat, which is expensive, is less likely to be wasted, and partly because food-waste processors discourage people from disposing of meat by that route because it may be hazardous to health.

The process is called hydrolysis because it requires water. The cleavage of the disaccharide sucrose by the enzyme invertase to give two monosaccharides, glucose and fructose, which was discussed in Section 2.2, is an example of hydrolysis, the overall equation for the reaction being:

$$C_{12}H_{22}O_{11} + H_2O \rightarrow 2 \times C_6H_{12}O_6$$

The cleavage of any of the bonds between glucose units in starch (Section 2.4.2) is also a hydrolytic reaction. The product of amylase activity on starch is maltose (also known as maltobiose), a disaccharide comprising two glucose units linked by an $\alpha(1{-}4)$ bond between the first carbon atom of one glucose unit and the fourth of the other. The glucose units in the linear chains in starch are linked by the same bonds. Cleavage of this bond by the enzyme maltase is also a hydrolytic reaction, and indeed it has exactly the same overall equation as the one given for the hydrolysis of sucrose above, since fructose and glucose are isomers.

Starch will be present in waste food, but a much more abundant polysaccharide in the feedstocks for anaerobic digestion is cellulose. This is because most animals, including humans, cannot digest cellulose, so it is present in animal waste (the exceptions being ruminants, which can digest cellulose because of the presence of symbiotic bacteria in their digestive system, and some termites); and because cellulose makes up most of the biomass of straw and other agricultural waste products. Cellulose is resistant to amylase activity because the glucose units are linked by $\beta(1{-}4)$ bonds; that is, bonds between the same carbon atoms as those that are linked in starch but in a different orientation to the $\alpha(1{-}4)$ bonds in starch. Hydrolytic bacteria can break down cellulose because they produce an enzyme called cellulase that will cleave the β(1–4) bonds in cellulose.

Like amylase activity on starch, the product of cellulase activity on cellulose is a disaccharide formed of two glucose units, with the chemical formula $C_{12}H_{22}O_{11}$; but the units are linked by β(1–4) bonds, making this disaccharide cellobiose rather than maltose. Maltose and cellobiose are shown in Figure 5.2. Cellobiose is

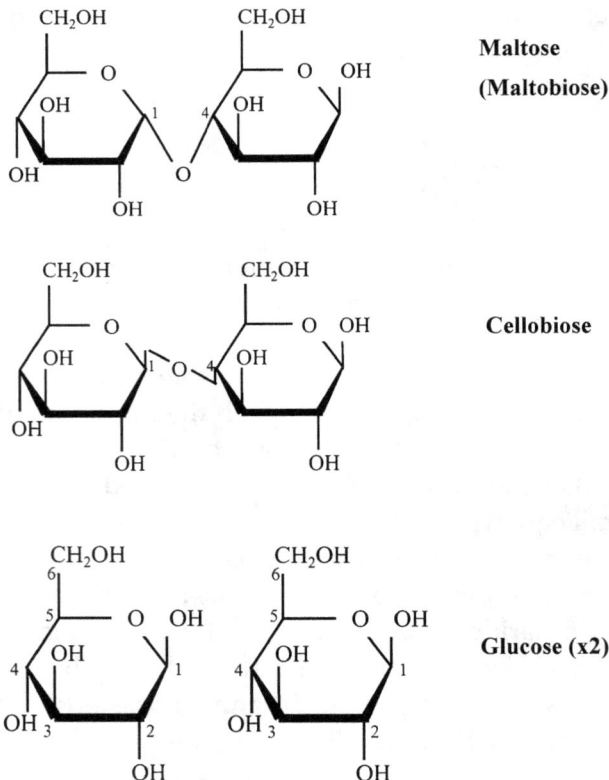

Figure 5.2   Diagrams showing the structures of the disaccharides, maltose (maltobiose) and cellobiose, and two glucose molecules, which are formed when the disaccharides undergo hydrolytic cleavage. The numbers indicate the position of the carbon atoms. Note that individual hydrogen atoms bonded to the carbon atoms are not shown.

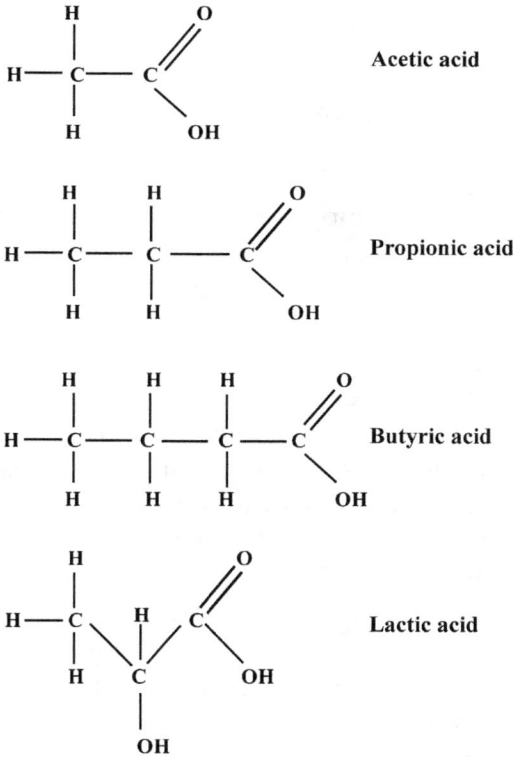

Figure 5.3 Diagrams showing the structures of the volatile organic acids produced in the acidogeneis stage of anaerobic digestion.

cleaved in another hydrolytic reaction by the enzyme cellobiase (β-glucosidase), to give two glucose units (Figure 5.2).

In the next stage, acidogenesis, the hydrolysed compounds are fermented by acidogenic bacteria to form organic acids, including acetic acid, propionic acid, butyric acid and lactic acid (Figure 5.3); as well as ethanol, methanol, ammonia (NH_3), hydrogen and carbon dioxide. In the following stage, acetogenesis, other bacteria convert the organic acids into acetic acid; this process also produces more ammonia, hydrogen and carbon dioxide. Finally, the methanogenic archaea convert the acetic acid and some of the carbon dioxide to methane, using the hydrogen produced in the preceding reactions as a reducing agent. This can be represented by the simple equations:

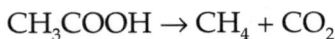

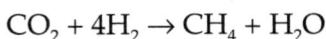

$$CH_3COOH \rightarrow CH_4 + CO_2$$
$$CO_2 + 4H_2 \rightarrow CH_4 + H_2O$$

The resulting crude biogas is typically made up of approximately 60% methane and 40% carbon dioxide, with traces of nitrogen, oxygen and hydrogen sulphide, and is saturated with water. Other problematic contaminants include siloxanes, which are silica-containing organic compounds characterised by a Si–O–Si linkage that are produced by digestion of materials found in soaps and detergents. One of these extraordinary molecules, hexamethyldisiloxane, is shown in Figure 5.4. If siloxanes are not removed, silicon dioxide (sand or glass) deposits may build up during combustion in the boiler, generator or engine in which the fuel is used.

As well as the biogas, a solid co-product is produced that can be used as a fertiliser or soil conditioner. Indeed, the solid matter produced from treated sewage has long been used as a fertiliser; the difference now is that value is also being ascribed to the sewage gas. Similarly, farmyard manure has previously been allowed to rot and then been spread over fields as a fertiliser, or simply spread over fields and allowed to break down *in situ* (muck-spreading). There is interest now in using manure as feedstock for anaerobic digestion. As well as the potential value of the gas, there is the advantage that the methane that is produced as the manure breaks down is captured and used rather than allowed to escape into the

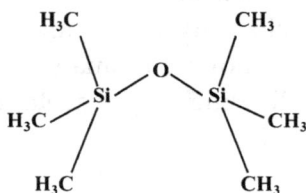

Figure 5.4 Diagram showing the structure of a siloxane molecule, hexamethyldisiloxane.

atmosphere, where it is a potent greenhouse gas with 86 times the global warming potential of carbon dioxide.

5.3 Biomethane (Renewable Natural Gas)

Crude biogas can be used as a low-grade fuel without further processing as long as the levels of hydrogen sulphide, siloxanes and other contaminants are not too high, and farm-scale operations will often use crude biogas to provide heating on-site and nothing more. However, if it is to be fed into the national gas grid or used to make compressed natural gas for use as a transport fuel (Section 1.2), crude biogas must first be upgraded to biomethane by removing the water vapour, carbon dioxide and contaminants. Biomethane is also known as renewable natural gas (RNG).

The first step is to dry the gas by passing it through one or more of a heat exchanger, a blower, a chiller to condense the water vapour out, or a dessicant. One possible dessicant is silica gel, a granular, porous form of silicon dioxide (SiO_2); passing the gas through a silica gel filter also removes siloxanes. Next, other contaminants must be removed, and there are several possible further purification steps that can be used. One is to filter the gas through activated carbon (also known as activated charcoal): that is, carbon, usually derived from charcoal, that has been processed so that it contains multiple small pores, giving it a huge, absorptive and reactive surface area. This removes both sulphides and siloxanes.

Activated carbon may also be used as the molecular sieve in a process called pressure swing absorption, as may other microporous substances such as zeolites (aluminosilicate minerals). Pressure swing adsorption exploits the tendency for gases to be attracted to solid surfaces at high pressure. Gases can be separated because different gases are attracted to different extents, depending on the nature of the material used as the trap. The biogas is filtered into the absorbing material at high pressure (>100 pounds per square inch gauge). As the pressure is released (the process 'swings' to low pressure), the methane escapes while the contaminating gases remain in the matrix.

Another possible purification process is to filter the gas through iron oxides, Fe_3O_4 (iron (II, III) oxide or ferrous-ferric oxide, also known as magnetite) and Fe_2O_3 (iron (III) oxide or ferric oxide; also known as haematite or maghemite, depending on the shape of the crystals). These oxides react with hydrogen sulphide to give iron sulphide (FeS_2; also known as iron pyrite or fool's gold) and water. An 'iron sponge' may be created with iron oxides in a matrix of wood chips.

An alternative or additional way of removing sulphides is biofiltration, in which microbes are used to oxidise the sulphides to sulphate (SO_4^{2-}). Typically, the gas is forced through a matrix of rock, gravel, coke, polyurethane foam, plastic beads or even sphagnum peat moss, on which a film of microbial slime is growing. Different designs have an assortment of names, including biofilter, trickling filter, bioscrubber and biotrickling filter.

Another alternative is the amine scrubber. This system has the advantage that it removes carbon dioxide as well as hydrogen sulphide and is commonly used in the petrochemical industry to remove these gases from liquid petroleum gas (Section 1.2). In that industry the process is known as gas sweetening because of the 'sour' odour associated with hydrogen sulphide. There is also interest in using amine scrubbing to remove carbon dioxide and pollutants from gases produced by burning fossil fuels, for example in coal-fired power stations. The process uses aqueous solutions of alkylamines (usually abbreviated to amines) such as monoethanolamine (C_2H_7NO), diethanolamine $C_4H_{11}NO_2$) and methyldiethanolamine $C_5H_{13}NO_2$) (Figure 5.5). For monoethanolamine, the reaction with hydrogen sulphide can be represented by the following equation, where ethanolamine is represented as RNH_2:

$$RNH_2 + H_2S \rightarrow RNH_3^+ + SH^-$$

For carbon dioxide, there are two steps, the first producing a Zwitter ion intermediate which then reacts with another molecule

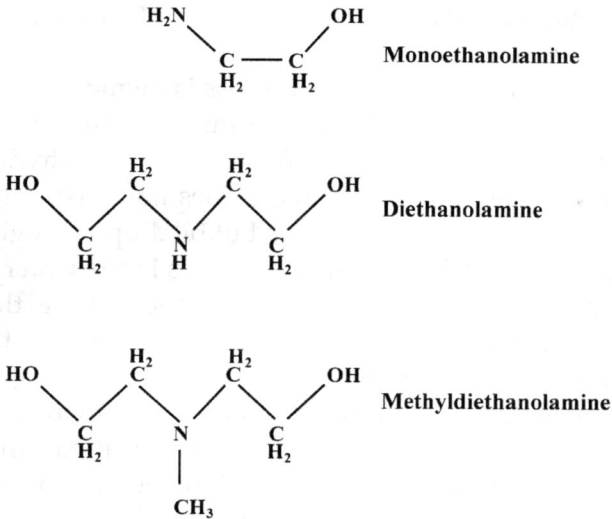

Figure 5.5   Diagrams showing the structures of a three alkylamines used in amine scrubbing of biogas.

of ethanolamine to produce the stable products of protonated ethanolamine and carbamate:

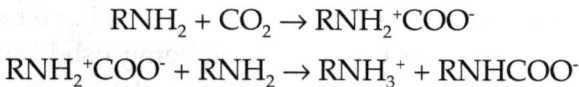

$$RNH_2 + CO_2 \rightarrow RNH_2^+COO^-$$

$$RNH_2^+COO^- + RNH_2 \rightarrow RNH_3^+ + RNHCOO^-$$

In the first part of the process, the biogas passes up through a solution of the amine and the carbon dioxide and hydrogen sulphide are absorbed. The clean gas is taken off the top of the absorber unit, while the amine solution containing the carbon dioxide and hydrogen sulphide (sometimes called rich amine) is taken off the bottom. The amine is then recycled by passing it through a stripper and reboiler which produces fresh (lean) amine solution to go back through the absorber and concentrated carbon dioxide and hydrogen sulphide gas. In oil refineries, most of this gas is hydrogen sulphide, and this is used to produce elemental sulphur, making up most of the more than 60 million tonnes of sulphur that is produced annually worldwide.

Yet another possible clean-up process is membrane separation, in which the gas is pumped through thin, hollow fibres made of membranes through which the carbon dioxide and hydrogen sulphide will permeate but the methane does not. Lastly there is the water wash process, in which gas is bubbled up through water at high pressure. The carbon dioxide dissolves in the water while the methane passes through and is removed. The water is then regenerated by transferring it to a low pressure vessel where the carbon dioxide comes out of solution.

The aim of these purification processes is to produce high-quality biomethane (renewable natural gas) that can then be fed into the national gas grid. High quality biomethane should contain 98% methane, less than 2% carbon dioxide and less than 4 parts per million hydrogen sulphide.

5.4 Feedstock

Almost any organic material can be used as a feedstock for anaerobic digestion, although some materials are more putrescible (readily decomposable) than others. There have been reports of grain and other food raw materials being used but this is an inappropriate use of food when other feedstocks are available. Much more acceptable is the use of waste products, including food waste, waste paper, grass cuttings, straw and other agricultural waste, including sugar cane and beet bagasse (Sections 2.3.2 and 2.3.3); industrial effluents, sewage and manure (note that woody material is not suitable without pre-treatment because of its lignin content: lignin is not broken down efficiently by anaerobic bacteria), and general household waste otherwise destined for landfill.

Biogas production also represents an alternative use for dedicated non-woody energy crops described in the previous chapter on biomass and for algae. These feedstocks may be used on their own or co-digested with manure, sewage or other waste products to increase the overall energy value of the feedstock and to facilitate the digestion process, in the same way that grass cuttings may be mixed with waste products to speed up digestion in a domestic composter.

5.5 Commercial Production

As with several sectors of the bioenergy industry already described in this book, the production of biogas by anaerobic digestion is still in its infancy as a commercial process and it remains to be seen how successful and widely adopted it will be. In the UK, it qualifies for many of the economic incentives provided by the UK government and described in Section 1.5, including Renewables Obligation Certificates, Feed-in Tariffs and Renewable Heat Incentives. Large-scale operations may be connected to the national gas grid, or use the biogas that is produced to generate electricity and feed that into the national electricity grid. Gas fed into the national grid would have to be biomethane (renewable natural gas) rather than crude biogas. To date there is little use of compressed natural gas as a vehicle fuel in the UK.

There are a number of different digester designs, one of which is shown in Figure 5.6. In this design, multiple large tanks are linked, with the digestate moving from one tank to the next, gas

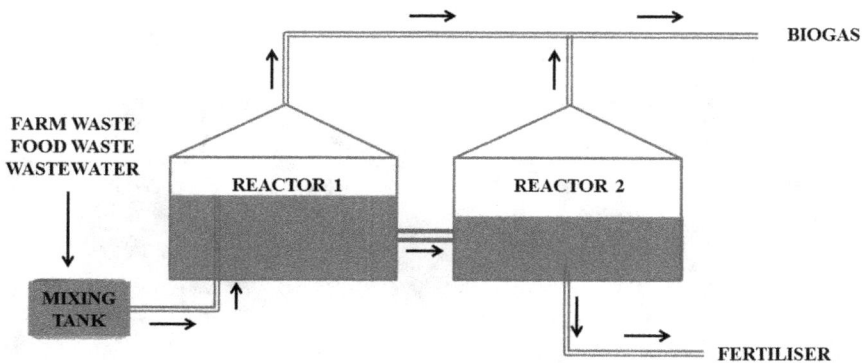

Figure 5.6   Diagram showing a typical design for a large, multi-tank anaerobic digestion plant.

being drawn off from each tank and the spent digestate being removed from the final tank to be dried and sold as fertiliser. The crude biogas from large operations like this would usually pass to a purification plant where CO_2, hydrogen sulphide and other contaminants would be removed before the gas is fed into the national gas grid. A plant of this type is shown in Figure 5.7. This operation

Figure 5.7   A biogas production plant in the UK that uses food waste as feed-stock.

is run by Biogen near Bedford in the UK and uses domestic food waste collected by local authorities. Most local authorities in the UK now require domestic food waste to be separated from general household waste so that it can be sold on; another example of a waste product that previously would have had a disposal cost associated with it, but now has a value. Another large anaerobic digestion operation was opened at Didcot Sewage Works in Oxfordshire in 2010, becoming the first anaerobic digester in the UK using sewage as the main feedstock to be connected to the national gas grid. The plant is owned jointly by Thames Water and British Gas and produces biomethane.

Smaller operations may use the gas that is generated to provide local heat or power, and there is already a market for farm-scale anaerobic digesters that use manure and other agricultural waste to provide gas for heating on the farm. One possible design for a farm-scale anaerobic digester is shown in Figure 5.8. In this design, the digester is below the surface of the ground, allowing the farmer to feed manure and other agricultural waste into an inlet tank from which it passes into the main digester. The digestate moves through the system as more waste is added to the inlet tank, eventually collecting in an

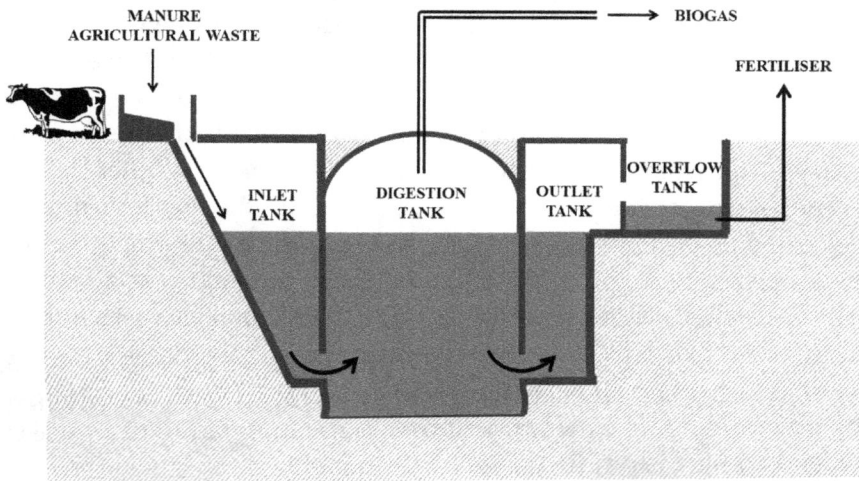

Figure 5.8 Diagram showing a typical design for a farm-scale anaerobic digester.

outlet tank from which spent slurry overflows into a collection tank where it can be removed and used as a soil improver and fertiliser. Biogas collects at the top of the main tank and is drawn off.

There are some problems that have to be overcome if a farm-scale digester is to work efficiently. One of these is that populations of anaerobic micro-organisms with the optimal make-up of bacteria and archaea take time to establish, and may be difficult to maintain. Establishment can be achieved by seeding a digester with partially digested material from a working digester, in the expectation that the right microbes will be present. However, the methanogenic archaea that produce methane require the pH to be between 6.5 and 8, which is quite a narrow range; and while large, dedicated operators may monitor conditions in the digester to ensure that methane production occurs efficiently, it is questionable whether farmers will have the time and expertise to do so for farm-scale operations.

As discussed above (Section 5.3), biogas also contains contaminants, including traces of hydrogen sulphide. Hydrogen sulphide gives an unpleasant smell (often described as being like rotten eggs) and is toxic, so its production and release is subject to regulatory control in most countries. The amount of hydrogen sulphide that is produced depends on the feedstock and can be reduced by adding iron chloride ($FeCl_2$) to the digestion. Nevertheless, the biogas may need 'scrubbing' before use, which requires the installation of additional equipment. A third problem is the disposal of wastewater from the digester. This contains, amongst other things, some of the ammonia that is produced in the process and is regarded as potentially polluting, so may require treatment before it can be released into watercourses. These problems are all solvable, but the solutions may require additional investment in equipment and close monitoring of the digestion process.

The rising cost of energy together with the incentives provided by government and the attraction of generating value from waste mean that biogas production is likely to increase rapidly in the next few years. The fact that waste materials make up most of the feed-stock means that the carbon intensity of biogas and biomethane is extremely low. Indeed, the California Air Resources Board has calculated a carbon intensity of 11.26 g/MJ for compressed natural gas produced from landfill gas, giving a greenhouse gas saving of 87% when used instead of petroleum-derived petrol.

At present the industry is at the early stages of becoming established in the United Kingdom. One of the issues that has inhibited its development is the strict control system that is applied to waste, and the United Kingdom government's Environment Agency has recently developed a 'Quality Protocol' to enable biogas to be produced and fed into the national gas grid without the producer having to comply with all of the waste regulatory controls. The aim is to stimulate growth in biomethane production to reach 1% of the country's demand for gas by 2020. Adoption of the protocol is being promoted through the Green Gas Certification Scheme run by the Renewable Energy Association.

The potential of biogas is also attracting interest in the United States, with landfill, wastewater, farm waste and food waste all seen as attractive sources. Indeed, the American Biogas Council lists over two thousand wastewater treatment plants that already have a biogas 'project' and 200 farms producing biogas, seven of which make biomethane.

6 CONCLUSIONS

This book describes the bioenergy industry: that is, the industry that produces energy from biological sources. The point is made several times in the introduction to the book (Section 1.1) that using biological material to produce energy is nothing new. When our ancestors used fire for warmth and to tenderise food, the fuel they burnt was wood; and the main domestic fuel, at least in Europe, right up to the industrial revolution was charcoal. Even fossil fuels, coal, petroleum oil and natural gas, have biological origins. However, a new industry has sprung up in the last few decades in response to the recognition that fossil fuels may be available in huge reserves but are nevertheless finite, and are being used at such a rate that atmospheric carbon dioxide levels are climbing rapidly to levels not experienced for tens of millions of years, with potentially catastrophic consequences for climate change.

Bioenergy is subdivided in this book into biofuel (liquid fuel primarily used for transport), biomass (solid material principally used for burning to produce electricity) and biogas (a gas containing methane that is produced by anaerobic digestion of biological (usually waste) material). Biofuel is further subdivided into bioethanol, which is a substitute for petrol (US gasoline), and biodiesel. These definitions are not universally accepted, and while the terms biofuel, biomass and biogas are widely used they are not always applied in the same way.

The first major modern bioenergy industry began in Brazil in the 1970s with the Brazilian Government's National Alcohol Program (Proálcool), an initiative that promoted the production of bioethanol from sugar cane as part replacement for petroleum-derived petrol (Section 2.3.2). The Brazilian government of the time should take great credit in showing that a bioenergy programme could be successful, and the global bioenergy industry that we have today was undoubtedly inspired by the success of the Proálcool programme.

One of the factors that drove the expansion of the Brazilian bioethanol industry was government support, and one of the conclusions to be drawn from the preceding chapters in this book is that almost the entire bioenergy industry is dependent on government subsidies and/or other political support of one sort or another. A purely economic case could possibly be made for the production of biodiesel from palm oil, but little else. Fossil fuels are still abundant and very cheap sources of energy (although consumers would probably disagree with that), and are extremely difficult to compete with. The energy market is also notoriously volatile, making long-term planning difficult and new ventures into the market without government protection very risky. Nevertheless, the degree of government support is striking and unusual, particularly in countries such as the US where governments are generally more inclined to let markets find their own way.

Another clear conclusion to be drawn from the preceding chapters is that bioenergy is not going to fuel the planet, because the energy demands of the modern world far outstrip potential bioenergy production. No doubt this has been recognised from the start within the industry and the government departments that are supporting it, with bioenergy being seen as one of the contributors to a diverse energy supply system that is less dependent on fossil fuels than it was, not as the complete solution to ensuring future energy security. Nevertheless, it is important to make that point clear.

A key question that follows this conclusion is whether or not it is worth compromising food supply to provide a small proportion of the world's energy. It should be remembered when considering this that the food versus fuel debate is a relatively recent one because until about 2008 food security was not regarded as a global problem; rather it was an issue that affected specific regions and populations due to one or more of poverty, political instability, poor infrastructure, population displacement and extreme weather events. Indeed, while diversifying energy production may have been the primary motive for politicians to promote the development of the bioenergy industry in Europe and the US at the turn of the century, a second motive was undoubtedly to create new markets for agricultural produce at a time when the price of food was so low that farmers were struggling to make ends meet on both sides of the Atlantic. In other words, commentators who blame the promotion of the bioenergy industry for increasing food prices should remember that an increase in food prices was initially one of the aims of many bioenergy programmes.

There is no doubt that food prices have been climbing steadily this century as a result of a change in the ratio of demand to supply, and that rising prices could be a portent of more severe problems in the future if the issue of meeting the increasing demand for agricultural products cannot be addressed. However, the use of agricultural products for bioenergy rather than food is only one of the factors leading to a rise in demand, with population growth and increasing *per capita* food consumption also important contributors. This was discussed in Section 1.6.

While food prices were already on the rise in the early years of the 21st century, the issue really came to wide attention in 2008 as prices spiked (Figure 1.3) on the back of a severe drought in Australia, a country that in most years is an important food exporter. Attitudes to global food security changed abruptly, and another extreme weather event followed in 2010, this time in Russia. Food security was now definitely on the agenda of governments and international agencies all around the world, and has remained there since, with the use of agricultural products for fuel rather than food coming under close and sometimes hostile scrutiny. A United Nations expert, Jean Ziegler, even called the practice of converting food crops into biofuel 'a crime against humanity'.

These issues are never simple, of course, and there are some facts about agriculture that are sometimes ignored in the debate. The first is that higher food prices may be bad for consumers but they are good for farmers; a balance must be achieved and maintained, because long-term food security is not helped by farmers going out of business. Second, farmers generally respond to rising prices by increasing production. There may be a limit to how far this can go, but it is notable that while 42% of US maize grain is currently being used for bioethanol rather than food, maize grain production in the US in 2013 was 40% higher than in 2000 (Section 2.4.3); so the demand for bioethanol from maize grain is being met almost entirely by additional production. Plant breeders have also responded to the increasing demand by investing heavily in maize breeding programmes, including biotech. There is also the possibility of bringing some of the millions of hectares of land that were abandoned in the US in the 1990s, due to low farm prices, back into agricultural production. This was mainly 'marginal' land, where production was affected by poor soil quality or adverse climatic conditions, but it may be viable now that prices have risen, or may be suitable for some of the novel biofuel crops that can tolerate sub-optimal growing conditions.

It is also important to note that now that the bioenergy industry is established it may invest in diversifying its feedstock. Bioethanol production, for example, may be based almost entirely on sugars and grain starch at present, but there is a strong push within the industry to exploit the cellulose and other complex polymers in straw and other agricultural waste products (Section 2.5). Similarly, the possibility of making biodiesel from second generation, non-food crops that will survive on land not suitable for food production is being explored (Section 3.5), as is the use of second generation biomass crops for electricity generation and biological waste products for biogas production. It is unlikely that investment in these programmes would have been made if the bioenergy industry were not well established. If the use of second generation feedstocks displaces feedstocks that can be used for food, the food versus fuel debate will become a food and fuel debate.

One way for the bioenergy industry to deflect criticism is to demonstrate that it provides a significant environmental benefit compared with the use of fossil fuels, in particular by reducing emissions of carbon dioxide and other greenhouse gases. Calculating the carbon intensity of a biofuel so that it can be compared with other energy sources is difficult, as was discussed in Section 1.4; but it is evident that many biofuels are associated with a greenhouse gas saving compared with fossil fuels, although the saving will never be 100%.

Perhaps the most obvious environmental benefit of the bioenergy industry is in the use of waste products to generate energy, something that is already a striking feature of the industry: from the firing of electricity generators with forestry waste to the production of biogas from food waste, manure and sewage. These waste products, some of which might otherwise find their way into landfill sites where they would decompose to produce methane that would end up in the atmosphere as a potent greenhouse gas, now have a value instead of a disposal cost associated with them; providing a welcome new income stream for the industries that produce them. While some sectors of the bioenergy industry may be controversial, the use of waste to provide energy and fuel does seem to be a win-win situation.

FURTHER READING
AND RESOURCES

The issue of greenhouse gas savings associated with biofuels is reviewed in detail in this paper:

Whitaker, J., Ludley, K.E., Rowe, R., Taylor, G. and Howard, D.C. (2010). Sources of variability in greenhouse gas and energy balances for biofuel production: A systematic review. *Global Change Biology and Bioenergy*, **2**, 99–112.

An explanation of the United Kingdom's Renewable Transport Fuel Obligation and the carbon intensity figures quoted in this book are given in this document:

Carbon and Sustainability Reporting Within the Renewable Transport Fuel Obligation Requirements and Guidance. Government Recommendation to the Office of the Renewable Fuels Agency. January 2008. Department for Transport, Great Minster House, 76 Marsham Street, London SW1P 4DR.

A very different estimate of the carbon intensity of ethanol from US maize is given in this report:

> Boland, S. and Unnasch, S. (2014). *Carbon Intensity of Marginal Petroleum and Corn Ethanol Fuels. Life Cycle Associates Report LCA.6075.83.2014*. Prepared for Renewable Fuels Association.

A detailed and in-depth consideration of the ethics of biofuels is given in the following document from the Nuffield Council on Bioethics:

> Nuffield Council on Bioethics (2011). *Biofuels: Ethical Issues*. Nuffield Press, Abingdon, Oxfordshire.

A negative view on the ethics of biofuels is provided by David Walker from the University of Sheffield, in this paper:

> Walker, D.A. (2010). Biofuels – For better or worse? *Annals of Applied Biology*, **156**, 319–327.

The latest IPCC assessment on climate change is available in this report:

> Intergovernmental Panel on Climate Change (2013). The Physical Science Basis. *Working Group I Contribution to the Fifth Assessment Report of the Intergovernmental Panel on Climate Change*. Cambridge University Press, Cambridge.

The issue of food security in the context of wheat breeding is reviewed in this paper:

> Curtis, T.Y and Halford, N.G. (2014). Food security: The challenge of increasing wheat yield and the importance of not compromising food safety. *Annals of Applied Biology*, **164**, 354–372.

Readers who are interested in genetically modified crops will get more information from this book:

Halford, N.G. (2012). *Genetically Modified Crops*, 2nd Edition. Imperial College Press, London.

Several excellent reviews of plant fatty acid and lipid biosynthesis have been written by Professor John Harwood of Cardiff University, including:

Harwood, J.L. (2010). *Plant Fatty Acid Synthesis*. The AOCS Lipid Library. Available online: http://lipidlibrary.aocs.org/plantbio/fa_biosynth/index.htm. Accessed 31 July 2014.

Yeast features many times in this book, particularly in Chapter 2 on bioethanol. The following is a comprehensive, high-level book on its biology:

Dickinson, J.R. and Schweizer, M. (eds) (2004). *Metabolism and Molecular Physiology of Saccharomyces cerevisiae, 2nd Edition*. Taylor and Francis, London and Philadelphia.

The present book contains data in tabular and graphical form from a variety of sources. Some of this was obtained directly from source, and some through the Index Mundi web site, http://www.indexmundi.com/, which would be a good place to start for readers seeking further statistics on crop production and prices.

INDEX

www.ingramcontent.com/pod-product-compliance
Lightning Source LLC
Chambersburg PA
CBHW061254220326
41599CB00028B/5643